Operator Theory
Advances and Applications
Vol. 96

Editor:
I. Gohberg

Schur Functions, Operator Colligations, and Reproducing Kernel Pontryagin Spaces

Daniel Alpay
Aad Dijksma
James Rovnyak
Hendrik de Snoo

Birkhäuser Verlag
Basel · Boston · Berlin

Authors:

Daniel Alpay
Department of Mathematics
Ben-Gurion University of the Negev
P.O.Box 653
84105 Beer-Sheva
Israel
e-mail: dany@ivory.bgu.ac.il

James Rovnyak
Department of Mathematics
University of Virginia
Charlottesville, VA 22903-3199
USA
e-mail: rovnyak@Virginia.EDU

Aad Dijksma
Department of Mathematics
University of Groningen
P.O.Box 800
9700 AV Groningen
The Netherlands
e-mail: dijksma@math.rug.nl

Hendrik de Snoo
Department of Mathematics
University of Groningen
P.O.Box 800
9700 AV Groningen
The Netherlands
e-mail: desnoo@math.rug.nl

1991 Mathematics Subject Classification 47A48, 47B50 (primary), 46C20, 46E22, 47A45 (secondary)

A CIP catalogue record for this book is available from the
Library of Congress, Washington D.C., USA

Deutsche Bibliothek Cataloging-in-Publication Data
**Schur functions, operator colligations and reproducing kernel
Pontryagin spaces** / Daniel Alpay ... – Basel ; Boston ; Berlin :
Birkhäuser, 1997
   (Operator theory ; Vol. 96)
   ISBN 3-7643-5763-0 (Basel ...)
   ISBN 0-8176-5763-0 (Boston ...)

© 1997 Birkhäuser Verlag, P.O. Box 133, CH-4010 Basel, Switzerland
Printed on acid-free paper produced from chlorine-free pulp. TCF ∞
Cover design: Heinz Hiltbrunner, Basel
Printed in Germany
ISBN 3-7643-5763-0
ISBN 0-8176-5763-0

9 8 7 6 5 4 3 2 1

Dedicated to Heinz Langer
on the occasion of his 60-th birthday,
in appreciation, admiration, and amity.

CONTENTS

CHAPTER 1

PONTRYAGIN SPACES AND OPERATOR COLLIGATIONS

After a review of reproducing kernel Pontryagin spaces, it is shown in §1.1 that a holomorphic kernel has the same number of negative squares for every region of analyticity. Background on colligations and their characteristic functions is presented in §1.2. Results from operator theory on Julia operators, the indices of a selfadjoint operator, and contractions are discussed in §1.3. An important result in §1.4 gives conditions that the closure of a linear relation is the graph of a continuous operator. In §1.5, the complementation properties of contractively contained spaces are used to show that, in natural situations, sums and differences of reproducing kernels in the indefinite case behave as in the nonnegative case, provided that suitable index conditions hold.

1.1 Reproducing kernel Pontryagin spaces

An **inner product space** is a pair $\left(\mathfrak{F}, \langle \cdot, \cdot \rangle_{\mathfrak{F}}\right)$ consisting of a linear space $\mathfrak{F}$ over the complex number field $\mathbf{C}$ and a mapping $\langle \cdot, \cdot \rangle_{\mathfrak{F}} : \mathfrak{F} \times \mathfrak{F} \to \mathbf{C}$ which is linear in the first variable and symmetric and called the **inner product**. We usually indicate the inner product space by referring to the linear space in the pair only, and we often use a subscript in the inner product of the pair to refer to the space. If $\mathfrak{F}$ is an inner product space, we call $\left(\mathfrak{F}, -\langle \cdot, \cdot \rangle_{\mathfrak{F}}\right)$ its **antispace** and denote it by $-\mathfrak{F}$. We define **Cartesian product** $\mathfrak{F} \times \mathfrak{G}$ and **orthogonal direct sum** $\mathfrak{F} \oplus \mathfrak{G}$ of inner product spaces $\mathfrak{F}$ and $\mathfrak{G}$ as in the case of Hilbert spaces.

By a **Kreĭn space** we mean an inner product space $\mathfrak{F}$ which can be written as the orthogonal direct sum $\mathfrak{F} = \mathfrak{F}_+ \oplus \mathfrak{F}_-$, where $\mathfrak{F}_+$ is a Hilbert space and $\mathfrak{F}_-$ is the antispace of a Hilbert space. Such a representation is called a **fundamental decomposition**. The associated **fundamental symmetry** or **signature operator** is the operator $J_{\mathfrak{F}}$ on $\mathfrak{F}$ defined by

$$J_{\mathfrak{F}}(f_+ + f_-) = f_+ - f_-$$

whenever $f_\pm \in \mathfrak{F}_\pm$. The numbers $\text{ind}_\pm \mathfrak{F} = \dim \mathfrak{F}_\pm$, which are either integers or ∞, are independent of the choice of fundamental decomposition and called the **positive** and **negative indices** of $\mathfrak{F}$. A **Pontryagin space** is a Kreĭn space with finite negative index (in some sources it is the positive index which is assumed to be finite, and results have to be converted). Hilbert spaces that appear in this work are assumed to be separable.

A topology is induced on a Kreĭn space $\mathfrak{F}$ by any fundamental decomposition $\mathfrak{F} = \mathfrak{F}_+ \oplus \mathfrak{F}_-$. First form the **associated Hilbert space** $|\mathfrak{F}| = \mathfrak{F}_+ \oplus |\mathfrak{F}_-|$ by replacing $\mathfrak{F}_-$ by its antispace $|\mathfrak{F}_-|$. The Hilbert space has an **associated norm** $\|\cdot\|$. Two norms arising from different fundamental decompositions can be shown to be equivalent. Therefore the norm topologies resulting from two fundamental decompositions are the same. All notions of continuity and convergence are understood to be with respect to this common topology, which is called the **strong topology** of $\mathfrak{H}$. The Riesz representation theorem and closed graph theorem for Kreĭn spaces follow immediately from their Hilbert space counterparts.

We write $\mathfrak{L}(\mathfrak{F})$ for the space of continuous (linear) operators on a Kreĭn space $\mathfrak{F}$, and $\mathfrak{L}(\mathfrak{F}, \mathfrak{G})$ for the continuous operators on $\mathfrak{F}$ into a second Kreĭn space $\mathfrak{G}$. If $A \in \mathfrak{L}(\mathfrak{F}, \mathfrak{G})$, its **adjoint** is the unique operator $A^* \in \mathfrak{L}(\mathfrak{G}, \mathfrak{F})$ such that

$$\langle Af, g \rangle_{\mathfrak{G}} = \langle f, A^* g \rangle_{\mathfrak{F}}, \qquad f \in \mathfrak{F}, \ g \in \mathfrak{G}.$$

At the same time, $A \in \mathfrak{L}(|\mathfrak{F}|, |\mathfrak{G}|)$ for any associated Hilbert spaces $|\mathfrak{F}|$ and $|\mathfrak{G}|$. If we write $A^\times \in \mathfrak{L}(|\mathfrak{G}|, |\mathfrak{F}|)$ for the Hilbert space adjoint, then $A^* = J_{\mathfrak{F}} A^\times J_{\mathfrak{G}}$, where $J_{\mathfrak{F}}$ and $J_{\mathfrak{G}}$ are the corresponding fundamental symmetries. In particular, $\|A^*\| = \|A^\times\| = \|A\|$, where $\|\cdot\|$ is the operator norm relative to the Hilbert spaces $|\mathfrak{F}|$ and $|\mathfrak{G}|$. The identity operator on a Kreĭn space $\mathfrak{F}$ is written 1 or sometimes $1_{\mathfrak{F}}$ to indicate the underlying space.

A **subspace** of a Kreĭn space is a linear manifold (which need not be closed). A subspace $\mathfrak{M}$ of a Kreĭn space $\mathfrak{H}$ is said to be **nonpositive** if $\langle f, f \rangle \leq 0$ for all $f \in \mathfrak{M}$, and **uniformly negative** if a number $\delta > 0$ exists such that

$$\langle f, f \rangle \leq -\delta \|f\|^2, \qquad f \in \mathfrak{M},$$

relative to some associated norm for $\mathfrak{H}$. The term "maximal" applied to either of these notions means a subspace which has the property and is not properly contained in another subspace with the same property. Similar definitions are made for **nonnegative** and **uniformly positive** subspaces. If $\mathfrak{H} = \mathfrak{H}_+ \oplus \mathfrak{H}_-$ is a fundamental decomposition with $\mathfrak{H}_+$ a Hilbert space and $\mathfrak{H}_-$ the antispace of a Hilbert space, then $\mathfrak{H}_-$ is maximal uniformly negative and $\mathfrak{H}_+$ is maximal uniformly positive. Every maximal uniformly negative (positive) subspace of $\mathfrak{H}$ arises in this way.

By a **regular subspace** of a Kreĭn space $\mathfrak{H}$ we mean a closed subspace $\mathfrak{M}$ which is itself a Kreĭn space in the inner product of $\mathfrak{H}$. We sometimes call such subspaces simply **Kreĭn subspaces**. A regular subspace which is a Hilbert space or a Pontryagin space in the inner product of $\mathfrak{H}$ is also called a **Hilbert subspace** or **Pontryagin subspace** of $\mathfrak{H}$. An operator $A \in \mathfrak{L}(\mathfrak{H})$ is **selfadjoint** if $A = A^*$. If $A, B \in \mathfrak{L}(\mathfrak{H})$ are two selfadjoint operators, $A \leq B$ means that

$$\langle Af, f \rangle_{\mathfrak{H}} \leq \langle Bf, f \rangle_{\mathfrak{H}}$$

for all $f \in \mathfrak{H}$. This relation is a partial ordering in the set of selfadjoint operators in $\mathfrak{L}(\mathfrak{H})$. An operator $P \in \mathfrak{L}(\mathfrak{H})$ is a **projection** if it is selfadjoint and $P^2 = P$. The class of regular subspaces of a Kreĭn space $\mathfrak{H}$ coincides with the class of subspaces of $\mathfrak{H}$ which are ranges of projections. It also coincides with the class of closed subspaces $\mathfrak{M}$ of $\mathfrak{H}$ for which $\mathfrak{H} = \mathfrak{M} \oplus \mathfrak{M}^{\perp}$. For an arbitrary subspace $\mathfrak{L}$ of $\mathfrak{H}$ (not necessarily regular), $\mathfrak{L}^{\perp} = \{f \in \mathfrak{H} : f \perp \mathfrak{L}\}$ is called the **orthogonal companion** of $\mathfrak{L}$. **Orthogonality** of two vectors f, g in $\mathfrak{H}$, $f \perp g$, means that $\langle f, g \rangle_{\mathfrak{H}} = 0$, and $f \perp \mathfrak{L}$ means $f \perp g$ for all $g \in \mathfrak{L}$. In the case that $\mathfrak{H} = \mathfrak{L} \oplus \mathfrak{L}^{\perp}$, that is, when $\mathfrak{L}$ is a regular subspace of $\mathfrak{H}$, we also call $\mathfrak{L}^{\perp}$ the **orthogonal complement** of $\mathfrak{L}$ and write $\mathfrak{L}^{\perp} = \mathfrak{H} \ominus \mathfrak{L}$. An example of a closed subspace for which the relation $\mathfrak{H} = \mathfrak{L} \oplus \mathfrak{L}^{\perp}$ fails is the one-dimensional subspace spanned by a nonzero **neutral vector**, that is, a vector f such that $\langle f, f \rangle_{\mathfrak{H}} = 0$ (in this example $\mathfrak{L} \subseteq \mathfrak{L}^{\perp}$, so the sum of $\mathfrak{L}$ and $\mathfrak{L}^{\perp}$ is not direct). The orthogonal direct sum of two Kreĭn spaces $\mathfrak{H}$ and $\mathfrak{K}$ is often written

$$\begin{pmatrix} \mathfrak{H} \\ \mathfrak{K} \end{pmatrix}.$$

This notation has an obvious extension to any number of summands.

An operator $A \in \mathfrak{L}(\mathfrak{H}, \mathfrak{K})$ is said to be

(1) **isometric** if $A^*A = 1$,
(2) **coisometric** if $AA^* = 1$,
(3) **unitary** if it is both isometric and coisometric, and
(4) **partially isometric** if $AA^*A = A$.

Such operators have properties much as in the Hilbert space case (Dritschel and Rovnyak [1996], Theorems 1.7, 1.8). For example, the range of an isometry is a Kreĭn subspace. It is important in these results that operators are everywhere defined and continuous. An **isomorphism** from a Kreĭn space $\mathfrak{H}$ to a Kreĭn space $\mathfrak{K}$ is a one-to-one and onto linear mapping that preserves inner products. Since the topologies of the spaces are determined by their inner products, such a mapping is automatically continuous. Thus the class of Kreĭn space isomorphisms from $\mathfrak{H}$ to $\mathfrak{K}$ coincides with the set of unitary operators in $\mathfrak{L}(\mathfrak{H}, \mathfrak{K})$. Equality of the indices of $\mathfrak{H}$ and $\mathfrak{K}$ is a necessary and sufficient condition for the existence of an isomorphism between the spaces.

Pontryagin spaces have properties which are not shared by all Kreĭn spaces. For example, a theorem of L. S. Pontryagin states that any dense subspace $\mathfrak{M}$ of a Pontryagin space contains a maximal uniformly negative subspace (Bognár [1974], Theorem 1.4, p. 185, and p. 207; Iokhvidov, Kreĭn, and Langer [1982], Lemma 2.1).

LEMMA 1.1.1. *Let $\mathfrak{H}$ be a Pontryagin space with negative index κ.*

(1) *The Gram matrix $G = \left(\langle g_j, g_i \rangle_{\mathfrak{H}} \right)_{i,j=1}^{n}$ of any finite number $g_1, \ldots, g_n$ of vectors in $\mathfrak{H}$ can have no more than κ negative eigenvalues.*

(2) *Every total set $\mathfrak{S}$ in $\mathfrak{H}$ contains a finite subset whose Gram matrix has exactly κ negative eigenvalues.*

Eigenvalues are counted according to multiplicity. In a matrix

$$\left(c_{ij} \right)_{i,j=1}^{n},$$

the term c_{ij} denotes the entry in the i-th row and j-th column.

Lemma 1.1.1 is an immediate consequence of the theorem of L. S. Pontryagin cited above and the following property of Gram matrices.

LEMMA 1.1.1'. *Let $g_1, \ldots, g_n$ be vectors in an inner product space $\mathfrak{H}$, and let $G = \left(\langle g_j, g_i \rangle_{\mathfrak{H}} \right)_{i,j=1}^{n}$ be their Gram matrix. Then the number of negative eigenvalues of G coincides with the maximum dimension of a subspace $\mathfrak{N}$ of the span of $g_1, \ldots, g_n$ which is the antispace of a Hilbert space in the inner product of $\mathfrak{H}$.*

Proof. Let $\lambda_1, \ldots, \lambda_n$ be the eigenvalues of G arranged in such a way that $\lambda_1 \leq \cdots \leq \lambda_n$. Let $\nu(G)$ be the number of negative eigenvalues of G as an operator on $\mathbf{C}^n$ (n-dimensional Euclidean space). If $\lambda_r < 0$ for some r (that is, if $r \leq \nu(G)$), then G has orthonormal eigenvectors

$$x_j = \begin{pmatrix} \gamma_{j1} \\ \vdots \\ \gamma_{jn} \end{pmatrix}, \qquad j = 1, \ldots, r,$$

corresponding to the negative eigenvalues $\lambda_1, \ldots, \lambda_r$. Put

$$u_1 = \gamma_{11}\, g_1 + \cdots + \gamma_{1n}\, g_n,$$
$$\cdots$$
$$u_r = \gamma_{r1}\, g_1 + \cdots + \gamma_{rn}\, g_n.$$

Then for all $i, j = 1, \ldots, r$,

$$\langle u_j, u_i \rangle_{\mathfrak{H}} = \sum_{p,q=1}^{n} \gamma_{jq} \bar{\gamma}_{ip} \langle g_q, g_p \rangle_{\mathfrak{H}} = \langle G x_j, x_i \rangle_{\mathbf{C}^n} = \lambda_j \delta_{ij},$$

where δ_{ij} is the Kronecker symbol. Therefore $u_1, \ldots, u_r$ span an r-dimensional subspace $\mathfrak{N}$ of the span of $g_1, \ldots, g_n$ which is the antispace of a Hilbert space in the inner product of $\mathfrak{H}$. Hence

$$\nu(G) \leq \max \dim \mathfrak{N},$$

where the maximum is over all such subspaces.

Suppose next that $\mathfrak{N}$ is an r-dimensional subspace of the span of $g_1, \ldots, g_n$ which is the antispace of a Hilbert space in the inner product of $\mathfrak{H}$. We show that $\lambda_r < 0$. Choose $v_1, \ldots, v_r$ in $\mathfrak{N}$ such that

$$\langle v_j, v_i \rangle_{\mathfrak{H}} = -\delta_{ij}, \qquad i, j = 1, \ldots, r.$$

Then

$$v_1 = \gamma_{11} g_1 + \cdots + \gamma_{1n} g_n,$$
$$\cdots$$
$$v_r = \gamma_{r1} g_1 + \cdots + \gamma_{rn} g_n,$$

for some numbers γ_{ip}, $i = 1, \ldots, r$, $p = 1, \ldots, n$. Define vectors $x_1, \ldots, x_r$ in $\mathbf{C}^n$ from these numbers as above. Calculating as before, we obtain

$$\langle G x_j, x_i \rangle_{\mathbf{C}^n} = \langle v_j, v_i \rangle_{\mathfrak{H}} = -\delta_{ij}, \qquad i, j = 1, \ldots, r.$$

Hence if $y = \eta_1 x_1 + \cdots + \eta_r x_r$ for some scalars $\eta_1, \ldots, \eta_r$, then

$$\langle G y, y \rangle_{\mathbf{C}^n} = -\sum_{j=1}^{r} |\eta_j|^2.$$

In particular, $y = 0$ only when $\eta_1 = \cdots = \eta_r = 0$. Therefore the span $\mathfrak{L}$ of $x_1, \ldots, x_r$ is an r-dimensional subspace of $\mathbf{C}^n$. We also obtain

$$\max \left\{ \langle G y, y \rangle_{\mathbf{C}^n} : y \in \mathfrak{L}, \ \|y\|_{\mathbf{C}^n} = 1 \right\} < 0,$$

since a continuous function on a compact set attains its maximum value. By the minimax principle (Halmos [1958], p. 181),

$$\lambda_r = \min \ \max \left\{ \langle G y, y \rangle_{\mathbf{C}^n} : y \in \mathfrak{M}_r, \ \|y\|_{\mathbf{C}^n} = 1 \right\},$$

where the minimum is over all subspaces $\mathfrak{M}_r$ of $\mathbf{C}^n$ of dimension r. Thus $\lambda_r < 0$, and so

$$\nu(G) \geq \max \dim \mathfrak{N},$$

where the maximum is over all subspaces $\mathfrak{N}$ of the span of $g_1, \ldots, g_n$ which are antispaces of Hilbert spaces in the inner product of $\mathfrak{H}$. Thus equality holds, which proves the result. $\qquad\qquad\square$

By a **kernel** is meant a function $K(w, z)$ on $\Omega \times \Omega$ with values in $\mathfrak{L}(\mathfrak{F})$ for some Kreĭn space $\mathfrak{F}$ and nonempty set Ω. We say that $K(w, z)$ is **Hermitian** if

$$K(w, z)^* = K(z, w), \quad w, z \in \Omega.$$

If Ω is an open set in the complex plane, a Hermitian kernel $K(w, z)$ on $\Omega \times \Omega$ is said to be **holomorphic** if it is holomorphic in z for each fixed w and holomorphic in $\bar{w}$ for each fixed z.

The symbol κ is reserved for a finite nonnegative integer. Let $K(w, z)$ be a Hermitian kernel on $\Omega \times \Omega$ with values in $\mathfrak{L}(\mathfrak{F})$ for some Kreĭn space $\mathfrak{F}$ and set Ω. We say that $K(w, z)$ has κ **negative squares**, in symbols

$$\mathrm{sq}_- K = \kappa,$$

if for any finite set of points $w_1, \ldots, w_n$ in Ω and vectors $f_1, \ldots, f_n$ in $\mathfrak{F}$, the Hermitian matrix

$$(1.1.1) \qquad\qquad \left(\langle K(w_j, w_i) f_j, f_i \rangle_{\mathfrak{F}} \right)_{i,j=1}^n$$

has at most κ negative eigenvalues, and at least one such matrix has exactly κ negative eigenvalues. As before, we count eigenvalues according to multiplicity. We say that $K(w, z)$ is **nonnegative** if the condition is satisfied with $\kappa = 0$, that is, the matrix (1.1.1) is nonnegative in every case. If the condition is not met for any nonnegative number κ, we write $\mathrm{sq}_- K = \infty$.

Let $\mathfrak{H}$ be a Pontryagin space whose elements are functions on a set Ω with values in a Kreĭn space $\mathfrak{F}$. We then call $\mathfrak{H}$ a **functional Pontryagin space** and $\mathfrak{F}$ a **coefficient space**. It is always assumed in such a situation that the linear operations of $\mathfrak{H}$ are defined pointwise, that is, for all $f, g \in \mathfrak{H}$ and $a, b \in \mathbf{C}$,

$$(af + bg)(w) = af(w) + bg(w)$$

for each $w \in \Omega$. Equivalently, for each $w \in \Omega$, the **evaluation mapping** $E(w)$ which takes f into $f(w)$ is linear from $\mathfrak{H}$ into $\mathfrak{F}$. A kernel $K(w, z)$ on $\Omega \times \Omega$ with values in $\mathfrak{L}(\mathfrak{F})$ is said to be a **reproducing kernel** for $\mathfrak{H}$ if for each $w \in \Omega$ and $f \in \mathfrak{F}$,

(1) $K(w, z)f$ belongs to $\mathfrak{H}$ as a function of z, and
(2) $\langle h(\cdot), K(w, \cdot)f \rangle_{\mathfrak{H}} = \langle h(w), f \rangle_{\mathfrak{F}}$ for every $h(\cdot)$ in $\mathfrak{H}$.

When such a function exists, we call $\mathfrak{H}$ a **reproducing kernel Pontryagin space**. In this case, the set of functions $K(w, \cdot)f$ with $w \in \Omega$ and $f \in \mathfrak{F}$ is total in $\mathfrak{H}$.

THEOREM 1.1.2. *Let $\mathfrak{H}$ be a Pontryagin space of functions defined on a set Ω with values in a Kreĭn space $\mathfrak{F}$. Then $\mathfrak{H}$ has a reproducing kernel if and only if all evaluation mappings $E(w)$, $w \in \Omega$, act continuously from $\mathfrak{H}$ into $\mathfrak{F}$. When a reproducing kernel exists, it is unique. It is a Hermitian kernel given by*

$$K(w, z) = E(z)E(w)^*, \qquad w, z \in \Omega,$$

and sq_ $K = \kappa$, where κ is the negative index of $\mathfrak{H}$. If Ω is an open set in the complex plane and the elements of $\mathfrak{H}$ are holomorphic functions, then $K(w, z)$ is a holomorphic kernel.

Proof. If $\mathfrak{H}$ has a reproducing kernel, continuity of the evaluation mappings follows by the closed graph theorem. Conversely, if the evaluation mappings are continuous, we easily check that $E(z)E(w)^*$ is a reproducing kernel for $\mathfrak{H}$. The proof of uniqueness is also easy, and we omit it.

To prove the assertion on the number of negative squares, observe that

$$\langle K(u, v)f, g \rangle_{\mathfrak{F}} = \langle K(u, \cdot)f, K(v, \cdot)g \rangle_{\mathfrak{H}}, \qquad u, v \in \Omega, \ f, g \in \mathfrak{F}.$$

Hence matrices of the form (1.1.1) are Gram matrices for elements of the total set in $\mathfrak{H}$ consisting of functions $K(u, \cdot)f$ with $u \in \Omega$ and $f \in \mathfrak{F}$. Thus by Lemma 1.1.1, no matrix of the form (1.1.1) can have more than κ negative eigenvalues, and at least one such matrix has exactly κ negative eigenvalues, that is, sq_ $K = \kappa$.

If the elements of $\mathfrak{H}$ are holomorphic vector-valued functions on an open set Ω in the complex plane, then $E(\cdot)$ is a holomorphic operator-valued function, and hence the reproducing kernel is holomorphic. $\square$

Two functional Pontryagin spaces are said to be **isometrically equal** if they coincide as vector spaces and have the same inner products. Uniqueness in the next result is in this sense.

THEOREM 1.1.3. *Let $K(w, z)$ be a Hermitian kernel on $\Omega \times \Omega$ with values in $\mathfrak{L}(\mathfrak{F})$ for some Kreĭn space $\mathfrak{F}$ and set Ω, and assume that $K(w, z)$ has a finite number of negative squares. Then there exists a unique Pontryagin space $\mathfrak{H}$ of $\mathfrak{F}$-valued functions on Ω with reproducing kernel $K(w, z)$. If Ω is an open set in the complex plane and $K(w, z)$ is a holomorphic kernel, then the elements of $\mathfrak{H}$ are holomorphic functions.*

Proof. Existence. We reduce this to the well-known case of nonnegative kernels. Let $\mathfrak{H}_0$ be the span of functions $K(w, \cdot)f$ with w in Ω and f in $\mathfrak{F}$. Define a linear and symmetric inner product on $\mathfrak{H}_0$ by requiring that

$$\left\langle \sum_{j=1}^{n} K(w_j, \cdot)f_j, \sum_{i=1}^{n} K(w_i, \cdot)f_i \right\rangle_{\mathfrak{H}_0} = \sum_{i,j=1}^{n} \langle K(w_j, w_i)f_j, f_i \rangle_{\mathfrak{F}}$$

whenever $w_1, \dots, w_n \in \Omega$ and $f_1, \dots, f_n \in \mathfrak{F}$. As in the nonnegative case, it is not hard to see that the inner product is well defined and

$$\langle h(\cdot), K(w, \cdot)f \rangle_{\mathfrak{H}_0} = \langle h(w), f \rangle_{\mathfrak{F}}$$

for all $h \in \mathfrak{H}_0$, $w \in \Omega$, and $f \in \mathfrak{F}$.

If $\mathrm{sq}_- K = \kappa$, there exist $\alpha_1, \dots, \alpha_n \in \Omega$ and $f_1, \dots, f_n \in \mathfrak{F}$ such that the Gram matrix

$$\left(\langle K(\alpha_j, \cdot)f_j, K(\alpha_i, \cdot)f_i \rangle_{\mathfrak{H}_0} \right)_{i,j=1}^{n} = \left(\langle K(\alpha_j, \alpha_i)f_j, f_i \rangle_{\mathfrak{F}} \right)_{i,j=1}^{n}$$

has κ negative eigenvalues. By Lemma 1.1.1', $\mathfrak{H}_0$ contains a κ-dimensional subspace $\mathfrak{H}_-$ which is the antispace of a Hilbert space, and no subspace of $\mathfrak{H}_0$ of higher dimension has this property. Let $u_1(\cdot), \dots, u_\kappa(\cdot)$ be elements of $\mathfrak{H}_-$ such that

$$\langle u_j(\cdot), u_i(\cdot) \rangle_{\mathfrak{H}_0} = -\delta_{ij}, \qquad i, j = 1, \dots, \kappa.$$

Then a reproducing kernel for $\mathfrak{H}_-$ is defined by

$$K_-(w, z)f = -\sum_{j=1}^{\kappa} \langle f, u_j(w) \rangle_{\mathfrak{F}} u_j(z)$$

for all $w, z \in \Omega$ and $f \in \mathfrak{F}$.

Put

$$K_+(w, z) = K(w, z) - K_-(w, z),$$

and let $\mathfrak{H}_{0+}$ be the span of functions $K_+(w, \cdot)f$ with w in Ω and f in $\mathfrak{F}$. This subspace is orthogonal to $\mathfrak{H}_-$ in $\mathfrak{H}_0$, since

$$\langle K_+(w, \cdot)f, u_l(\cdot) \rangle_{\mathfrak{H}_0} = \langle f, u_l(w) \rangle_{\mathfrak{F}} + \sum_{j=1}^{\kappa} \langle f, u_j(w) \rangle_{\mathfrak{F}} \langle u_j(\cdot), u_l(\cdot) \rangle_{\mathfrak{H}_0} = 0$$

for all $l = 1, \dots, \kappa$. If

$$f(\cdot) = \sum_{j=1}^{n} K_+(w_j, \cdot)f_j, \qquad g(\cdot) = \sum_{i=1}^{n} K_+(w_i, \cdot)g_i$$

with $w_1, \ldots, w_n$ in Ω and $f_1, \ldots, f_n, g_1, \ldots, g_n$ in $\mathfrak{F}$, then

$$\langle f(\cdot), g(\cdot) \rangle_{\mathfrak{H}_0} = \left\langle \sum_{j=1}^{n} K(w_j, \cdot) f_j - \sum_{j=1}^{n} K_-(w_j, \cdot) f_j, \right.$$

$$\left. \sum_{i=1}^{n} K(w_i, \cdot) g_i - \sum_{i=1}^{n} K_-(w_i, \cdot) g_i \right\rangle_{\mathfrak{H}_0}$$

$$= \sum_{i,j=1}^{n} \langle K(w_j, w_i) f_j, g_i \rangle_{\mathfrak{F}} - \sum_{i,j=1}^{n} \langle K_-(w_j, w_i) f_j, g_i \rangle_{\mathfrak{F}}$$

$$- \sum_{i,j=1}^{n} \langle K_-(w_j, w_i) f_j, g_i \rangle_{\mathfrak{F}} + \sum_{i,j=1}^{n} \langle K_-(w_j, w_i) f_j, g_i \rangle_{\mathfrak{F}}$$

$$= \sum_{i,j=1}^{n} \langle K_+(w_j, w_i) f_j, g_i \rangle_{\mathfrak{F}}.$$

We use this to show that the kernel $K_+(w, z)$ is nonnegative. Argue by contradiction. If $K_+(w, z)$ is not nonnegative, we can find $\beta_1, \ldots, \beta_q$ in Ω and $\psi_1, \ldots, \psi_q$ in $\mathfrak{F}$ such that the element

$$u_{\kappa+1}(\cdot) = \sum_{j=1}^{q} K_+(\beta_j, \cdot) \psi_j$$

of $\mathfrak{H}_{0+}$ satisfies

$$\langle u_{\kappa+1}(\cdot), u_{\kappa+1}(\cdot) \rangle_{\mathfrak{H}_0} = \sum_{i,j=1}^{q} \langle K_+(\beta_j, \beta_i) \psi_j, \psi_i \rangle_{\mathfrak{F}} < 0.$$

Then the span of $\mathfrak{H}_-$ and $u_{\kappa+1}(\cdot)$ is a $(\kappa+1)$-dimensional subspace of $\mathfrak{H}_0$ which is the antispace of a Hilbert space. As noted above, this contradicts our hypothesis that $\mathrm{sq}_- K = \kappa$. Therefore $K_+(w, z)$ is nonnegative.

Appealing to the known nonnegative case, we conclude that $\mathfrak{H}_{0+}$ has a completion to a Hilbert space $\mathfrak{H}_+$ of functions on Ω with reproducing kernel $K_+(w, z)$, and the elements of $\mathfrak{H}_+$ are holomorphic if Ω is an open set in the complex plane and the kernel is holomorphic. Define a Pontryagin space $\mathfrak{H}$ such that $\mathfrak{H}_\pm$ are isometrically contained in $\mathfrak{H}$ and $\mathfrak{H} = \mathfrak{H}_+ \oplus \mathfrak{H}_-$. Since $\mathfrak{H}_\pm$ have reproducing kernels $K_\pm(w, z)$, $\mathfrak{H}$ has reproducing kernel

$$K(w, z) = K_+(w, z) + K_-(w, z).$$

This completes the existence part of the proof.

Uniqueness. Let $\mathfrak{K}$ be any Pontryagin space of functions on Ω with reproducing kernel $K(w, z)$. Then $\mathfrak{K}$ isometrically contains the spaces $\mathfrak{H}_0, \mathfrak{H}_{0+}, \mathfrak{H}_-$ constructed above, and $\mathfrak{H}_0$ is dense in $\mathfrak{K}$. It follows that $\mathfrak{H}_{0+}$ is dense in $\mathfrak{K} \ominus \mathfrak{H}_-$ (an element of $\mathfrak{K} \ominus \mathfrak{H}_-$ which is orthogonal to $\mathfrak{H}_{0+}$ would have to be orthogonal to $\mathfrak{H}_0 = \mathfrak{H}_{0+} + \mathfrak{H}_-$ and hence identically zero), and so $\mathfrak{K} \ominus \mathfrak{H}_-$ coincides with the space $\mathfrak{H}_+$ constructed above. Therefore $\mathfrak{K}$ is isometrically equal to the space $\mathfrak{H}$ constructed in the existence part of the proof. $\qquad\square$

THEOREM 1.1.4. *Let $K(w, z)$ be a holomorphic Hermitian kernel on $\Omega \times \Omega$ with values in $\mathfrak{L}(\mathfrak{F})$ for some Kreĭn space $\mathfrak{F}$ and region Ω. If the restriction $K_0(w, z)$ to $\Omega_0 \times \Omega_0$ has κ negative squares for some subregion Ω_0 of Ω, then $K(w, z)$ has κ negative squares on $\Omega \times \Omega$.*

In other words, for a holomorphic Hermitian kernel, the property of having κ negative squares is propagated to arbitrarily large regions in which the kernel is holomorphic. Connectivity of Ω is essential for this phenomenon. This is implied in the term **region**, by which we mean a nonempty open connected set in the complex plane.

The proof of Theorem 1.1.4 will be based on two lemmas.

LEMMA 1.1.5. *Let $K(w, z)$ be a holomorphic Hermitian kernel defined for $|w| < R$, $|z| < R$ with values in $\mathfrak{L}(\mathfrak{F})$ for some Kreĭn space $\mathfrak{F}$ and positive number R. If the restriction of $K(w, z)$ to $|w| < r$, $|z| < r$ is nonnegative for some $r \in (0, R)$, then $K(w, z)$ is nonnegative for $|w| < R$, $|z| < R$.*

LEMMA 1.1.6. *Let $K(w, z)$ be a holomorphic Hermitian kernel on $\Omega \times \Omega$ with values in $\mathfrak{L}(\mathfrak{F})$ for some Kreĭn space $\mathfrak{F}$ and region Ω. Suppose that $\Omega = \Omega_1 \cup \Omega_2$, where Ω_1 and Ω_2 are regions and the restrictions $K_1(w, z)$ and $K_2(w, z)$ of $K(w, z)$ to $\Omega_1 \times \Omega_1$ and $\Omega_2 \times \Omega_2$ are nonnegative. Then $K(w, z)$ is nonnegative on $\Omega \times \Omega$.*

Proof of Lemma 1.1.5. If $K(w, z) = \sum_{m,n=0}^{\infty} C_{mn} z^m \bar{w}^n$, $|w| < R$, $|z| < R$, it will be shown that $K(w, z)$ is nonnegative if and only if

$$(1.1.2) \qquad \left(C_{mn} \right)_{m,n=0}^{N} \geq 0, \qquad N \geq 0.$$

Since this characterization of nonnegativity does not depend on R, the lemma follows.

First assume that $K(w, z)$ is nonnegative for $|w| < R$, $|z| < R$. By considering

$$(1.1.3) \qquad K(\rho w, \rho z) = \sum_{m,n=0}^{\infty} C_{mn} \rho^{m+n} z^m \bar{w}^n,$$

where $0 < \rho < R$, we can reduce the proof of (1.1.2) to the case $R > 1$. In this case, by Cauchy's theorem

$$K(w, z) = \int_\Gamma \int_\Gamma \frac{K(u, v)}{(1 - \bar{w}u)(1 - z\bar{v})} \, d\sigma(u) \, d\sigma(v), \qquad |w| < 1, |z| < 1,$$

where σ is normalized Lebesgue measure on the unit circle $\Gamma = \{z : |z| = 1\}$. For any points $w_1, \ldots, w_M$ of modulus less than one and $f_1, \ldots, f_M$ in $\mathfrak{F}$, the nonnegativity of $K(w, z)$ implies that

$$0 \leq \sum_{i,j=1}^M \langle K(w_j, w_i)f_j, f_i \rangle_{\mathfrak{F}}$$

$$= \sum_{i,j=1}^M \int_\Gamma \int_\Gamma \frac{\langle K(u, v)f_j, f_i \rangle_{\mathfrak{F}}}{(1 - \bar{w}_j u)(1 - w_i \bar{v})} \, d\sigma(u) \, d\sigma(v)$$

$$= \int_\Gamma \int_\Gamma \langle K(u, v)\varphi(u), \varphi(v) \rangle_{\mathfrak{F}} \, d\sigma(u) \, d\sigma(v),$$

where $\varphi(z) = \sum_{j=1}^M f_j/(1 - \bar{w}_j z)$. By an approximation argument,

(1.1.4) $$\int_\Gamma \int_\Gamma \langle K(u, v)p(u), p(v) \rangle_{\mathfrak{F}} \, d\sigma(u) \, d\sigma(v) \geq 0$$

for every polynomial $p(z) = \sum_{j=0}^N g_j z^j$ with coefficients in $\mathfrak{F}$. Then

$$0 \leq \sum_{i,j=1}^N \int_\Gamma \int_\Gamma \langle K(u, v)g_j, g_i \rangle_{\mathfrak{F}} u^j \bar{v}^i \, d\sigma(u) \, d\sigma(v) = \sum_{i,j=1}^N \langle C_{ij}g_j, g_i \rangle_{\mathfrak{F}}.$$

Therefore (1.1.2) holds

Conversely, assume that (1.1.2) holds. Fix $f_1, \ldots, f_M$ in $\mathfrak{F}$. Again by considering (1.1.3), we can reduce the proof of the inequality

$$\sum_{i,j=1}^M \langle K(w_j, w_i)f_j, f_i \rangle_{\mathfrak{F}} \geq 0$$

to the case $R > 1$ and $w_1, \ldots, w_M$ of modulus less than one. Let $\varphi_N(z)$ be the N-th degree Taylor polynomial for the function $\varphi(z)$ defined above. Reversing the preceding steps, we use the hypothesis (1.1.2) to conclude that (1.1.4) holds with $p(z) = \varphi_N(z)$ for every $N \geq 0$. Passing to the limit as $N \to \infty$, we obtain

$$\sum_{i,j=1}^M \langle K(w_j, w_i)f_j, f_i \rangle_{\mathfrak{F}} = \int_\Gamma \int_\Gamma \langle K(u, v)\varphi(u), \varphi(v) \rangle_{\mathfrak{F}} \, d\sigma(u) \, d\sigma(v) \geq 0,$$

yielding the result. $\qquad \square$

Proof of Lemma 1.1.6. The intersection $\Omega_0 = \Omega_1 \cap \Omega_2$ is nonempty since $\Omega = \Omega_1 \cup \Omega_2$ is a region and hence connected. Our assumptions imply that the restriction $K_0(w, z)$ of $K(w, z)$ to $\Omega_0 \times \Omega_0$ is nonnegative. Let $\mathfrak{H}_0, \mathfrak{H}_1, \mathfrak{H}_2$ be the Hilbert spaces of holomorphic functions on $\Omega_0, \Omega_1, \Omega_2$ with reproducing kernels $K_0(w, z), K_1(w, z), K_2(w, z)$. For each $j = 1, 2$, the restriction mapping

$$R_j \varphi_j = \varphi_0, \quad \varphi_0 = \varphi_j|_{\Omega_0}, \quad \varphi_j \in \mathfrak{H}_j,$$

is a Hilbert space isomorphism from $\mathfrak{H}_j$ onto $\mathfrak{H}_0$.

Define a new Hilbert space $\mathfrak{H}$ in the following way. Each $\varphi_0 \in \mathfrak{H}_0$ has the form

$$\varphi_0 = \varphi_1|_{\Omega_0} = \varphi_2|_{\Omega_0}$$

for unique $\varphi_1 \in \mathfrak{H}_1$ and $\varphi_2 \in \mathfrak{H}_2$. Hence $\varphi_0 = \varphi|_{\Omega_0}$, where φ is a holomorphic $\mathfrak{F}$-valued function on Ω. Let $\mathfrak{H}$ be the space of all such functions φ. Then $\mathfrak{H}$ is a Hilbert space in a unique inner product such that the mapping $\varphi_0 \to \varphi$ is an isomorphism from $\mathfrak{H}_0$ onto $\mathfrak{H}$.

Evaluation mappings on $\mathfrak{H}$ are continuous. For if $w \in \Omega$, either $w \in \Omega_1$ or $w \in \Omega_2$. For definiteness, suppose that $w \in \Omega_1$. Choose a norm $\|\cdot\|_{\mathfrak{F}}$ that determines the strong topology of $\mathfrak{F}$. Let $\varphi \in \mathfrak{H}$, and put $\varphi_0 = \varphi|_{\Omega_0}$ and $\varphi_1 = \varphi|_{\Omega_1}$. Then $\varphi_0 \in \mathfrak{H}_0$, $\varphi_1 \in \mathfrak{H}_1$, and for any $f \in \mathfrak{F}$,

$$\left| \langle \varphi(w), f \rangle_{\mathfrak{F}} \right| = \left| \langle \varphi_1(w), f \rangle_{\mathfrak{F}} \right|$$
$$\leq M_w \|f\|_{\mathfrak{F}} \|\varphi_1\|_{\mathfrak{H}_1} = M_w \|f\|_{\mathfrak{F}} \|\varphi_0\|_{\mathfrak{H}_0} = M_w \|f\|_{\mathfrak{F}} \|\varphi\|_{\mathfrak{H}},$$

where M_w is a constant. Therefore $\mathfrak{H}$ has a reproducing kernel $L(w, z)$, which is nonnegative because $\mathfrak{H}$ is a Hilbert space.

By the definition of $\mathfrak{H}$, restriction to Ω_0 is an isomorphism from $\mathfrak{H}$ onto $\mathfrak{H}_0$. Therefore the restriction of $L(w, z)$ to $\Omega_0 \times \Omega_0$ is a reproducing kernel for $\mathfrak{H}_0$. Thus $K(w, z)$ and $L(w, z)$ are holomorphic Hermitian kernels which agree on $\Omega_0 \times \Omega_0$ and hence coincide identically. Therefore $K(w, z)$ is nonnegative.   □

Proof of Theorem 1.1.4. Suppose that $K(w, z)$ is nonnegative on Ω_0. Let $\Delta_0, \ldots, \Delta_n$ be disks in Ω such that the center of Δ_0 is in Ω_0 and the center of Δ_j is in $\Delta_{j-1}, j = 1, \ldots, n$. By Lemmas 1.1.5 and 1.1.6, $K(w, z)$ is nonnegative on Δ_0, Δ_1, and $\Delta_0 \cup \Delta_1$. In the same way, $K(w, z)$ is nonnegative on Δ_2 and $(\Delta_0 \cup \Delta_1) \cup \Delta_2$. Continuing in this way, we see that $K(w, z)$ is nonnegative on $\Delta_0 \cup \Delta_1 \cup \cdots \cup \Delta_n$. The result follows in the case $\kappa = 0$ because any finite set of points in Ω are contained in the union of a system of such disks.

For the general case, as in the proof of Theorem 1.1.3, we may write

$$K_0(w, z) = K_{0+}(w, z) + K_{0-}(w, z), \qquad w, z \in \Omega_0,$$

These notions are useful in the study of an arbitrary operator $T \in \mathfrak{L}(\mathfrak{H}, \mathfrak{K})$. Underlying spaces here are assumed to be Kreĭn spaces. By the Bognár-Krámli factorization, it is always possible to write

$$(1.3.3) \qquad\qquad 1_{\mathfrak{H}} - T^*T = \tilde{D}\tilde{D}^*,$$

where $\tilde{D} \in \mathfrak{L}(\tilde{\mathfrak{D}}, \mathfrak{H})$ is an operator such that ker $\tilde{D} = \{0\}$. We call an operator $\tilde{D}$ with these properties a **defect operator** for T and the domain space $\tilde{\mathfrak{D}}$ a **defect space**. In this situation,

$$\text{ind}_{\pm}(1_{\mathfrak{H}} - T^*T) = \text{ind}_{\pm} \tilde{\mathfrak{D}}.$$

The identity (1.3.3) expresses the fact that the operator matrix

$$(1.3.4) \qquad\qquad \begin{pmatrix} T \\ \tilde{D}^* \end{pmatrix} \in \mathfrak{L}(\mathfrak{H}, \mathfrak{K} \oplus \tilde{\mathfrak{D}})$$

is an isometric extension of T.

By a **Julia operator** for $T \in \mathfrak{L}(\mathfrak{H}, \mathfrak{K})$ we mean a unitary operator

$$(1.3.5) \qquad\qquad U = \begin{pmatrix} T & D \\ \tilde{D}^* & -L^* \end{pmatrix} : \begin{pmatrix} \mathfrak{H} \\ \mathfrak{D} \end{pmatrix} \to \begin{pmatrix} \mathfrak{K} \\ \tilde{\mathfrak{D}} \end{pmatrix}$$

in which $\mathfrak{D}$ and $\tilde{\mathfrak{D}}$ are Kreĭn spaces and $D \in \mathfrak{L}(\mathfrak{D}, \mathfrak{K})$ and $\tilde{D} \in \mathfrak{L}(\tilde{\mathfrak{D}}, \mathfrak{H})$ are operators with zero kernel. By Theorem 1.3.1, either one of the conditions on these kernels implies the other. The operator L belongs to $\mathfrak{L}(\tilde{\mathfrak{D}}, \mathfrak{D})$ and has no other condition imposed upon it. The identities

$$(1.3.6) \qquad\qquad \begin{cases} T^*T + \tilde{D}\tilde{D}^* = 1_{\mathfrak{H}}, \\ D^*D + LL^* = 1_{\mathfrak{D}}, \\ TT^* + DD^* = 1_{\mathfrak{K}}, \\ \tilde{D}^*\tilde{D} + L^*L = 1_{\tilde{\mathfrak{D}}}, \end{cases}$$

together with

$$(1.3.7) \qquad\qquad \begin{cases} T^*D = \tilde{D}L^*, \\ T\tilde{D} = DL, \end{cases}$$

express the unitarity of U. There is also redundancy in the conditions (1.3.6) and (1.3.7); see Brodskiĭ [1978], pp. 161–162 (English version) for the Hilbert space case, and Dritschel and Rovnyak, [1990], proof of Theorem B3, for the Kreĭn space case. In particular, $\tilde{D}$ is a defect operator for T, and D is a defect operator for T^*. The indices of the defect spaces are determined by T from the relations

$$(1.3.8) \qquad \begin{cases} \text{ind}_\pm(1_{\mathfrak{H}} - T^*T) = \text{ind}_\pm \tilde{\mathfrak{D}}, \\ \text{ind}_\pm(1_{\mathfrak{K}} - TT^*) = \text{ind}_\pm \mathfrak{D}. \end{cases}$$

If U is a Julia operator for T as given in (1.3.5), then U^* is a Julia operator for T^*. A Julia operator for any given operator $T \in \mathfrak{L}(\mathfrak{H}, \mathfrak{K})$ always exists. Moreover, if $T \in \mathfrak{L}(\mathfrak{H}, \mathfrak{K})$, where $\mathfrak{H}$ and $\mathfrak{K}$ are Kreĭn spaces, and if $\tilde{D} \in \mathfrak{L}(\tilde{\mathfrak{D}}, \mathfrak{H})$ is any defect operator for T, there exists a Julia operator (1.3.5) for T in which the operator $\tilde{D}$ is the given defect operator.

A Julia operator (1.3.5) is a unitary colligation $(\mathfrak{H}, \mathfrak{D}, \tilde{\mathfrak{D}}, U)$ in the case that $\mathfrak{H} = \mathfrak{K}$. We call such a unitary colligation a **Julia colligation**.

THEOREM 1.3.2. *Let $\mathfrak{H}, \mathfrak{F}, \mathfrak{G}$ be Kreĭn spaces, and let*

$$V = \begin{pmatrix} T & F \\ G & H \end{pmatrix} : \begin{pmatrix} \mathfrak{H} \\ \mathfrak{F} \end{pmatrix} \to \begin{pmatrix} \mathfrak{H} \\ \mathfrak{G} \end{pmatrix}$$

be any colligation. Then:

(1) *V is closely inner connected if and only if the only closed subspace $\mathfrak{M}$ of $\mathfrak{H}$ such that $V^*\mathfrak{M} \subseteq \mathfrak{M}$ is $\mathfrak{M} = \{0\}$;*

(2) *V is closely outer connected if and only if the only closed subspace $\mathfrak{M}$ of $\mathfrak{H}$ such that $V\mathfrak{M} \subseteq \mathfrak{M}$ is $\mathfrak{M} = \{0\}$;*

(3) *if V is closely connected, then the only closed subspace $\mathfrak{M}$ of $\mathfrak{H}$ such that $V\mathfrak{M} \subseteq \mathfrak{M}$ and $V^*\mathfrak{M} \subseteq \mathfrak{M}$ is $\mathfrak{M} = \{0\}$, and the converse is true if V is unitary.*

If V is a Julia colligation, then in addition:

(4) *V is closely inner connected if and only if the only closed subspace $\mathfrak{M}$ of $\mathfrak{H}$ such that $T^*\mathfrak{M} \subseteq \mathfrak{M}$ and $(1_{\mathfrak{H}} - TT^*)\mathfrak{M} = \{0\}$ is $\mathfrak{M} = \{0\}$;*

(5) *V is closely outer connected if and only if the only closed subspace $\mathfrak{M}$ of $\mathfrak{H}$ such that $T\mathfrak{M} \subseteq \mathfrak{M}$ and $(1_{\mathfrak{H}} - T^*T)\mathfrak{M} = \{0\}$ is $\mathfrak{M} = \{0\}$;*

(6) *V is closely connected if and only if the only closed subspace $\mathfrak{M}$ of $\mathfrak{H}$ such that $T\mathfrak{M} = \mathfrak{M}$ and $(1_{\mathfrak{H}} - T^*T)\mathfrak{M} = \{0\}$ is $\mathfrak{M} = \{0\}$.*

The closed subspaces $\mathfrak{M}$ which appear in (1)–(6) need not be regular.

Proof. (1) Assume that V is closely inner connected, and let $\mathfrak{M}$ be a closed subspace of $\mathfrak{H}$ such that $V^*\mathfrak{M} \subseteq \mathfrak{M}$. Then $T^*\mathfrak{M} \subseteq \mathfrak{M}$ and $F^*\mathfrak{M} = \{0\}$. Hence

$$F^*T^{*m}\mathfrak{M} = \{0\} \qquad \text{and} \qquad \mathfrak{M} \perp \text{ran } T^m F$$

for all $m \geq 0$. It follows that $\mathfrak{M} = \{0\}$, and so the condition is necessary. Sufficiency follows from the easily verified fact that the orthogonal companion of

$$\overline{\text{span}} \ \{\text{ran} \ T^m F : m \geq 0\}$$

in $\mathfrak{H}$ is a closed subspace $\mathfrak{M}$ such that $V^* \mathfrak{M} \subseteq \mathfrak{M}$.

(2) Apply (1) to the adjoint of V.

(3) Assume that V is closely connected, and let $\mathfrak{M}$ be a closed subspace of $\mathfrak{H}$ such that $V\mathfrak{M} \subseteq \mathfrak{M}$ and $V^*\mathfrak{M} \subseteq \mathfrak{M}$. Then $T\mathfrak{M} \subseteq \mathfrak{M}$, $G\mathfrak{M} = \{0\}$, $T^*\mathfrak{M} \subseteq \mathfrak{M}$, and $F^*\mathfrak{M} = \{0\}$. Therefore

$$\mathfrak{M} \perp \text{ran} \ T^{*n}G^*, \qquad \mathfrak{M} \perp \text{ran} \ T^m F$$

for all $m, n \geq 0$. Since V is closely connected, $\mathfrak{M} = \{0\}$.

Conversely, assume that V is unitary and the condition holds. To show that V is closely connected it is sufficient to show that the orthogonal companion $\mathfrak{M}$ of

$$\overline{\text{span}} \ \{\text{ran} \ T^m F, \ \text{ran} \ T^{*n}G^* : m, n \geq 0\}$$

in $\mathfrak{H}$ is invariant under V and V^*, since then the condition implies that $\mathfrak{M} = \{0\}$. By the definition of $\mathfrak{M}$,

$$F^* T^{*m} \mathfrak{M} = \{0\}, \qquad G T^n \mathfrak{M} = \{0\}, \qquad m, n \geq 0.$$

We show that

$$F^* T^{*m} T \mathfrak{M} = \{0\}, \qquad G T^n T^* \mathfrak{M} = \{0\}, \qquad m, n \geq 0.$$

By the unitarity of V, $F^*T = -H^*G$ and $T^*T = 1_{\mathfrak{H}} - G^*G$. For the case $m = 0$ in the first relation, we have $F^*T\mathfrak{M} = (-H^*G)\mathfrak{M} = \{0\}$. When $m > 0$,

$$F^* T^{*m} T h = F^* T^{*m-1} h - F^* T^{*m-1} G^* G h = 0, \qquad h \in \mathfrak{M},$$

establishing $F^* T^{*m} T \mathfrak{M} = \{0\}$ for all $m \geq 0$. The second relation holds because what has been proved for V applies as well to V^*. The two relations immediately yield $V\mathfrak{M} \subseteq \mathfrak{M}$ and $V^*\mathfrak{M} \subseteq \mathfrak{M}$, and hence $\mathfrak{M} = \{0\}$ by the condition. Thus V is closely connected.

(4)–(6) Assume now that V is a Julia colligation. This means that V is unitary and the operators F and G^* are one-to-one. For $i = 1, \ldots, 6$, let $C(i)$ be the condition on a subspace $\mathfrak{M}$ of $\mathfrak{H}$ in part (i) of the theorem. We prove (4)–(6) by showing the equivalences of the conditions in these parts with the corresponding conditions in (1)–(3).

$C(1) \Leftrightarrow C(4)$ Assume $C(1)$. Let $\mathfrak{M}$ be any closed subspace of $\mathfrak{H}$ such that $T^*\mathfrak{M} \subseteq \mathfrak{M}$ and $(1_{\mathfrak{H}} - TT^*)\mathfrak{M} = \{0\}$. Then

$$FF^*\mathfrak{M} = (1_{\mathfrak{H}} - TT^*)\mathfrak{M} = \{0\}.$$

Since F is one-to-one, $F^*\mathfrak{M} = \{0\}$. Thus for any $h \in \mathfrak{M}$, $V^*h = T^*h \in \mathfrak{M}$, and so $V^*\mathfrak{M} \subseteq \mathfrak{M}$. Hence $\mathfrak{M} = \{0\}$ by $C(1)$. Therefore C(4) holds.

Assume C(4). Let $\mathfrak{M}$ be any closed subspace of $\mathfrak{H}$ such that $V^*\mathfrak{M} \subseteq \mathfrak{M}$. As in the proof of (1), this implies that $T^*\mathfrak{M} \subseteq \mathfrak{M}$ and $F^*\mathfrak{M} = \{0\}$. Then

$$(1_{\mathfrak{H}} - TT^*)\mathfrak{M} = FF^*\mathfrak{M} = \{0\},$$

and hence $\mathfrak{M} = \{0\}$ by C(4). Therefore C(1) holds.

$C(2) \Leftrightarrow C(5)$ This follows from what we just showed applied to V^*.

$C(3) \Leftrightarrow C(6)$ Assume C(3). Let $\mathfrak{M}$ be any closed subspace of $\mathfrak{H}$ such that $T\mathfrak{M} = \mathfrak{M}$ and $(1_{\mathfrak{H}} - T^*T)\mathfrak{M} = \{0\}$. Then

$$G^*G\mathfrak{M} = (1_{\mathfrak{H}} - T^*T)\mathfrak{M} = \{0\},$$

and since G^* is one-to-one, $G\mathfrak{M} = \{0\}$. Hence for any $h \in \mathfrak{M}$, $Vh = Th$ and $V\mathfrak{M} = \mathfrak{M}$. Since V is unitary, we also have $\mathfrak{M} = V^*\mathfrak{M}$ in this case. Therefore $\mathfrak{M} = \{0\}$ by C(3). Thus C(6) follows.

Assume C(6). Let $\mathfrak{M}$ be any closed subspace of $\mathfrak{H}$ such that $V\mathfrak{M} \subseteq \mathfrak{M}$ and $V^*\mathfrak{M} \subseteq \mathfrak{M}$. As in the proof of (3), this implies that $T\mathfrak{M} \subseteq \mathfrak{M}$, $G\mathfrak{M} = \{0\}$, $T^*\mathfrak{M} \subseteq \mathfrak{M}$, and $F^*\mathfrak{M} = \{0\}$. Since V is unitary, V maps $\mathfrak{M}$ onto itself. The restriction of V to $\mathfrak{M}$ coincides with T, and so $T\mathfrak{M} = \mathfrak{M}$. Since

$$(1_{\mathfrak{H}} - T^*T)\mathfrak{M} = G^*G\mathfrak{M} = \{0\},$$

$\mathfrak{M} = \{0\}$ by C(6). Thus C(3) follows. $\qquad \square$

Two Julia operators (1.3.5) and

$$\begin{pmatrix} T & D' \\ \tilde{D}'^* & -L'^* \end{pmatrix} \in \mathfrak{L}(\mathfrak{H} \oplus \mathfrak{D}', \mathfrak{K} \oplus \tilde{\mathfrak{D}}')$$

are considered **essentially identical** if there exist Kreĭn space isomorphisms

(1.3.9) $$\varphi : \mathfrak{D} \to \mathfrak{D}' \quad \text{and} \quad \psi : \tilde{\mathfrak{D}} \to \tilde{\mathfrak{D}}'$$

such that

(1.3.10) $$D = D'\varphi, \qquad \tilde{D} = \tilde{D}'\psi, \qquad \varphi L = L'\psi.$$

THEOREM 1.5.7. *Let $\mathfrak{H}_1$ be a Pontryagin space of functions defined on a set Ω with values in a Kreĭn space $\mathfrak{F}$ and having reproducing kernel $K_1(w, z)$. Suppose that*

$$K_2(w, z) = A(z)K_1(w, z)A(w)^*, \qquad w, z \in \Omega,$$

where $A(z)$ is a function on Ω with values in $\mathfrak{L}(\mathfrak{F}, \mathfrak{G})$ for some Kreĭn space $\mathfrak{G}$. Then $\mathrm{sq}_-\, K_2 \leq \mathrm{sq}_-\, K_1$ and $K_2(w, z)$ is the reproducing kernel for a Pontryagin space $\mathfrak{H}_2$ of functions on Ω with values in $\mathfrak{G}$. The following assertions are equivalent:

(1) *multiplication by $A(\cdot)$, that is, the mapping $h(\cdot) \to A(\cdot)h(\cdot)$, acts as a continuous partial isometry from $\mathfrak{H}_1$ onto $\mathfrak{H}_2$ whose kernel is a Hilbert space;*

(2) *the set of elements $h(\cdot)$ in $\mathfrak{H}_1$ such that $A(\cdot)h(\cdot) \equiv 0$ is a Hilbert subspace of $\mathfrak{H}_1$;*

(3) $\mathrm{sq}_-\, K_2 = \mathrm{sq}_-\, K_1$.

Proof. It is clear that $\mathrm{sq}_-\, K_2 \leq \mathrm{sq}_-\, K_1$, and so $K_2(w, z)$ is the reproducing kernel for a Pontryagin space $\mathfrak{H}_2$ of functions on Ω with values in $\mathfrak{G}$.

$(1) \Leftrightarrow (2)$ If (1) holds, then (2) follows because the set of elements $h(\cdot)$ in $\mathfrak{H}_1$ such that $A(\cdot)h(\cdot) \equiv 0$ is the kernel of the partial isometry in (1).

Conversely, suppose that (2) holds. Then the set of elements $h(\cdot)$ in $\mathfrak{H}_1$ such that $A(\cdot)h(\cdot) \equiv 0$ is a closed subspace of $\mathfrak{H}_1$ and a Hilbert space in the inner product of $\mathfrak{H}_1$. Hence we may define a Pontryagin space $\mathfrak{P}$ as the set of functions $A(\cdot)h(\cdot)$ with $h(\cdot)$ in $\mathfrak{H}_1$, considered in the inner product which makes multiplication by $A(\cdot)$ a continuous partial isometry from $\mathfrak{H}_1$ onto $\mathfrak{P}$. It is not hard to see that $\mathfrak{P}$ has reproducing kernel $K_2(w, z) = A(z)K_1(w, z)A(w)^*$. In fact, if $w \in \Omega$ and $g \in \mathfrak{G}$, then

$$A(\cdot)K_1(w, \cdot)A(w)^*g$$

belongs to $\mathfrak{P}$ by the definition of the space. Any element of $\mathfrak{P}$ can be written in the form $k(\cdot) = A(\cdot)h(\cdot)$ with $h(\cdot)$ orthogonal to the kernel of the partial isometry, and so

$$\langle k(\cdot), A(\cdot)K_1(w, \cdot)A(w)^*g\rangle_{\mathfrak{P}} = \langle h(\cdot), K_1(w, \cdot)A(w)^*g\rangle_{\mathfrak{H}_1} = \langle k(w), g\rangle_{\mathfrak{G}}.$$

Therefore $\mathfrak{P}$ coincides with $\mathfrak{H}_2$ by Theorem 1.1.3. Thus (1) follows.

$(1) \Leftrightarrow (3)$ Assuming (1), we conclude that $\mathfrak{H}_1$ and $\mathfrak{H}_2$ have the same negative index, and hence (3) holds by Theorem 1.1.2.

Conversely, assume that (3) holds, or what is the same thing, that $\mathfrak{H}_1$ and $\mathfrak{H}_2$ have the same negative index. Let $\mathbf{R}$ be the linear relation in $\mathfrak{H}_2 \times \mathfrak{H}_1$ consisting of the span of all pairs

$$\big(K_2(w,\cdot)A(w)^*g, K_1(w,\cdot)A(w)^*g\big), \qquad w \in \Omega,\ g \in \mathfrak{G}.$$

It is easy to see that the relation is an isometry, and the domain of $\mathbf{R}$ is dense in $\mathfrak{H}_2$. By Theorem 1.4.2, the closure of the relation is the graph of a continuous everywhere defined isometry W^* from $\mathfrak{H}_2$ into $\mathfrak{H}_1$. Since $\mathfrak{H}_1$ and $\mathfrak{H}_2$ have the same negative index, the orthogonal complement of the range of W^*, that is, the kernel of W, is a Hilbert space. If $k = Wh$ for some $h \in \mathfrak{H}_1$ and $k \in \mathfrak{H}_2$, then

$$\begin{aligned}
\langle k(w), g \rangle_{\mathfrak{G}} &= \langle (Wh)(\cdot), A(\cdot)K_1(w,\cdot)A(w)^*g \rangle_{\mathfrak{H}_2} \\
&= \langle h(\cdot), K_1(w,\cdot)A(w)^*g \rangle_{\mathfrak{H}_1} = \langle A(w)h(w), g \rangle_{\mathfrak{G}}
\end{aligned}$$

for all $w \in \Omega$ and $g \in \mathfrak{G}$. Hence W is multiplication by $A(\cdot)$, and (1) follows. $\square$

It asserts that the colligations (2.4.2) and (2.4.3) are equivalent in the sense of (2.1.7) by means of an isomorphism between state spaces of a special form.

Proof. Since V^* is isometric on $\mathfrak{H} \oplus \mathfrak{G}$ to $\mathfrak{H} \oplus \mathfrak{F}$, and $\mathfrak{H} \oplus \mathfrak{G}$, $\mathfrak{H} \oplus \mathfrak{F}$, ran V^* are Pontryagin spaces having the same negative index,

$$\mathfrak{H} \oplus \mathfrak{F} = \operatorname{ran} V^* \oplus \mathfrak{C},$$

where $\mathfrak{C}$ is a Hilbert space. Put $\mathfrak{H}_0 = \mathfrak{C} \oplus \mathfrak{C} \oplus \cdots$. Define U_0 by (2.4.2), where

(2.4.5)
$$\begin{cases} A_0^* \, (c_0, c_1, c_2, \dots) = (c_1, c_2, c_3, \dots), \\ \begin{pmatrix} B_0^* \\ C_0^* \end{pmatrix} (c_0, c_1, c_2, \dots) = c_0, \end{cases}$$

for any $(c_0, c_1, c_2, \dots)$ in $\mathfrak{H}_0$. By construction, U_0^* is isometric and onto, and hence U_0 is unitary. Write

$$U_0 = \begin{pmatrix} A_1 & B_1 \\ C_1 & D_1 \end{pmatrix} : \begin{pmatrix} \mathfrak{H}_1 \\ \mathfrak{F} \end{pmatrix} \rightarrow \begin{pmatrix} \mathfrak{H}_1 \\ \mathfrak{G} \end{pmatrix},$$

where A_1, B_1, C_1, D_1 are the blocks in (2.4.2) and $\mathfrak{H}_1 = \mathfrak{H}_0 \oplus \mathfrak{H}$. Then

$$\begin{aligned}
\Theta_{U_0}(z) &= D_1 + zC_1(1 - zA_1)^{-1}B_1 \\
&= H + z \begin{pmatrix} 0 & G \end{pmatrix} \begin{pmatrix} (1 - zA_0)^{-1} & z(1 - zA_0)^{-1}B_0(1 - zT)^{-1} \\ 0 & (1 - zT)^{-1} \end{pmatrix} \begin{pmatrix} C_0 \\ F \end{pmatrix} \\
&= H + zG(1 - zT)^{-1}F \\
&= \Theta_V(z) \\
&= S(z)
\end{aligned}$$

for any z in $\Omega(S) \cap \Omega(\Theta_{U_0})$.

To verify that U_0 is closely connected, we must show that the ranges of the operators $(1 - wA_1)^{-1}B_1$ and

(2.4.6)
$$(1 - zA_1^*)^{-1}C_1^* = \begin{pmatrix} 0 \\ (1 - zT^*)^{-1}G^* \end{pmatrix},$$

where z and w range over a neighborhood of the origin, span a dense subspace of $\mathfrak{H}_1$. Let $\mathfrak{M}$ be the orthogonal companion of the span of these ranges in $\mathfrak{H}_1 = \mathfrak{H}_0 \oplus \mathfrak{H}$. It is the same thing to show that $\mathfrak{M} = \{0\}$. An argument in the proof of Theorem 1.3.2(3) shows that $U_0\mathfrak{M} \subseteq \mathfrak{M}$ and $U_0^*\mathfrak{M} \subseteq \mathfrak{M}$. Since V is closely outer connected, on inspection of (2.4.6) we see that $\mathfrak{M} \subseteq \mathfrak{H}_0$ because the ranges of

$(1 - zT^*)^{-1}G^*$ span a dense subspace of $\mathfrak{H}$. In particular, $\mathfrak{M}$ is a Hilbert space in the inner product of $\mathfrak{H}_1$. The restriction of U_0 to $\mathfrak{M}$ is thus a unitary operator on the Hilbert space $\mathfrak{M}$. Since

$$U_0|_{\mathfrak{M}} = A_0|_{\mathfrak{M}}$$

and the powers of A_0^* tend strongly to zero, $\mathfrak{M} = \{0\}$, as was to be shown.

Let (2.4.3) be another closely connected unitary colligation such that $\mathfrak{H}_0'$ is a Hilbert space and $S(z) = \Theta_{U_0'}(z)$ on a neighborhood of the origin. By Theorem 2.1.3(3), there is an isomorphism

$$W : \begin{pmatrix} \mathfrak{H}_0 \\ \mathfrak{H} \end{pmatrix} \to \begin{pmatrix} \mathfrak{H}_0' \\ \mathfrak{H} \end{pmatrix}$$

such that

$$\begin{pmatrix} W & 0 \\ 0 & 1_{\mathfrak{G}} \end{pmatrix} U_0 = U_0' \begin{pmatrix} W & 0 \\ 0 & 1_{\mathfrak{F}} \end{pmatrix}.$$

We show that W is diagonal:

$$W = \begin{pmatrix} W_0 & 0 \\ 0 & 1_{\mathfrak{H}} \end{pmatrix}.$$

It is enough to show that the restriction of W to $\mathfrak{H}$ is the identity operator. According to the construction of W in the proof of Theorem 2.1.3,

$$W\left(\left(1 - \bar{w}\begin{pmatrix} A_0 & B_0 \\ 0 & T \end{pmatrix}^*\right)^{-1}\begin{pmatrix} 0 \\ G^* \end{pmatrix}, \left(1 - \bar{w}\begin{pmatrix} A_0 & B_0 \\ 0 & T \end{pmatrix}\right)^{-1}\begin{pmatrix} C_0 \\ F \end{pmatrix}\right)\begin{pmatrix} g \\ f \end{pmatrix}$$

$$= \left(\left(1 - \bar{w}\begin{pmatrix} A_0' & B_0' \\ 0 & T \end{pmatrix}^*\right)^{-1}\begin{pmatrix} 0 \\ G^* \end{pmatrix}, \left(1 - \bar{w}\begin{pmatrix} A_0' & B_0' \\ 0 & T \end{pmatrix}\right)^{-1}\begin{pmatrix} C_0' \\ F \end{pmatrix}\right)\begin{pmatrix} g \\ f \end{pmatrix},$$

for all $f \in \mathfrak{F}$, $g \in \mathfrak{G}$, and w in a neighborhood of the origin. With $f = 0$, we get

$$W(1 - \bar{w}T^*)^{-1}G^*g = (1 - \bar{w}T^*)^{-1}G^*g.$$

Since V is closely outer connected, the restriction of W to $\mathfrak{H}$ is the identity operator.

The last statement in the theorem is clear for the choice of colligation (2.4.2) constructed in the proof. By what we just showed, (2.4.2) is unique except for replacement of the space $\mathfrak{H}_0$ by an isomorphic copy, and hence the statement holds for any choice.   □

In applications of Theorem 2.4.1, we shall be interested in detail which is encoded in the unitary operator U_0. For example, various conditions may be given which imply that V is unitary.

THEOREM 2.4.2. *Assume that $S(z) = \Theta_V(z)$ as in Theorem 2.4.1, and let the unitary operator (2.4.2) be chosen as in that result.*

(1) *The operator $(\begin{matrix} B_0 & C_0 \end{matrix})$ is partially isometric with initial space* ker V *and final space* ran $(1 - A_0 A_0^*)$. *Furthermore,*

$$(2.4.7) \qquad\qquad 1 - V^*V = \begin{pmatrix} B_0^* \\ C_0^* \end{pmatrix} (\begin{matrix} B_0 & C_0 \end{matrix})$$

and

$$(2.4.8) \qquad \begin{cases} \ker V = \operatorname{ran} \begin{pmatrix} B_0^* \\ C_0^* \end{pmatrix}, & \operatorname{ran} V^* = \ker (\begin{matrix} B_0 & C_0 \end{matrix}), \\[2mm] \ker B_0 = \operatorname{ran} V^* \cap \mathfrak{H}, & \ker C_0 = \operatorname{ran} V^* \cap \mathfrak{F}. \end{cases}$$

(2) *The identity*

$$(2.4.9) \qquad\qquad S(z)C_0^* = -G(1 - zT)^{-1}B_0^*$$

holds on a neighborhood of the origin, and ker $C_0^* \subseteq$ ker B_0^*.

(3) *If $\mathfrak{M} = \{f : S(z)f \equiv 0\}$, then $\mathfrak{M} = \operatorname{ran} C_0^*|_{\ker B_0^*}$. Each of the following conditions implies that V is unitary:*

(i) $C_0 = 0$;
(ii) $\mathfrak{M} = \{0\}$ *and $B_0 = 0$.*

In this case, $\mathfrak{H}_0 = \{0\}$ and $U_0 = V$.

It sometimes happens that $B_0 = 0$. Then by part (1) of the theorem, $C_0^*C_0$ is a projection such that ran $C_0^*C_0 = \operatorname{ran} C_0^* = \ker V$ is the subspace $\mathfrak{M}$ in part (3).

Proof. (1) The identity (2.4.7) follows from $U_0^*U_0 = 1$. Since the range of $(\begin{matrix} B_0 & C_0 \end{matrix})$ is in the Hilbert space $\mathfrak{H}_0$, the kernel of (2.4.7) is the same as the kernel of $(\begin{matrix} B_0 & C_0 \end{matrix})$. It follows that $(\begin{matrix} B_0 & C_0 \end{matrix})$ is a partial isometry (Dritschel and Rovnyak [1990], Theorem 1.1.6). In fact, the operator V^*V is the projection on the range of V^*, and so $1 - V^*V$ is the projection on (ran $V^*)^\perp = \ker V$, which is the initial space for $(\begin{matrix} B_0 & C_0 \end{matrix})$. Since

$$B_0 B_0^* + C_0 C_0^* = 1 - A_0 A_0^*$$

by the identity $U_0 U_0^* = 1$, the final space of $(\begin{matrix} B_0 & C_0 \end{matrix})$ is ran $(1 - A_0 A_0^*)$. The remaining assertions in (1) are now straightforward.

(2) From $U_0 U_0^* = 1$, we get $TB_0^* + FC_0^* = 0$ and $GB_0^* + HC_0^* = 0$, and hence

$$S(z)C_0^* = \Theta_V(z)C_0^* = HC_0^* + zG(1 - zT)^{-1}FC_0^*$$
$$= -GB_0^* - zG(1 - zT)^{-1}TB_0^* = -G(1 - zT)^{-1}B_0^*,$$

which is (2.4.9). In particular, $C_0^* h = 0$ implies $G(1 - zT)^{-1}B_0^* h \equiv 0$, and hence $B_0^* h = 0$ since V is closely outer connected.

(3) Since V is closely outer connected, $S(z)f \equiv 0$ if and only if $Hf = 0$ and $Ff = 0$. By the characterization of the kernel of V in (2.4.8), this holds if and only if

$$\begin{pmatrix} 0 \\ f \end{pmatrix} = \begin{pmatrix} B_0^* u \\ C_0^* u \end{pmatrix}$$

for some u in $\mathfrak{H}_0$, or, equivalently, $f \in \operatorname{ran} C_0^*|_{\ker B_0^*}$.

By (1), V is unitary if and only if $(\, B_0 \quad C_0 \,) = 0$. Assume (i). By (2), $B_0 = 0$ and so V is unitary. The same conclusion holds if we assume (ii): then $\operatorname{ran} C_0^* \subseteq \mathfrak{M}$ by (2.4.9), so the assumption that $\mathfrak{M} = \{0\}$ implies that $C_0 = 0$, and again V is unitary. Since U_0 is closely connected, it is only possible for V to be unitary when $\mathfrak{H}_0 = \{0\}$. $\square$

THEOREM 2.4.3. *Assume that $\mathfrak{F}$ and $\mathfrak{G}$ are Pontryagin spaces satisfying* $\operatorname{ind}_- \mathfrak{F} = \operatorname{ind}_- \mathfrak{G}$. *Let $\tilde{\mathfrak{H}}$ be a Pontryagin space, and let*

$$(2.4.10) \qquad \tilde{V} = \begin{pmatrix} \tilde{T} & \tilde{F} \\ \tilde{G} & \tilde{H} \end{pmatrix} : \begin{pmatrix} \tilde{\mathfrak{H}} \\ \mathfrak{F} \end{pmatrix} \to \begin{pmatrix} \tilde{\mathfrak{H}} \\ \mathfrak{G} \end{pmatrix}$$

be an isometric closely inner connected colligation. Suppose that $S(z) = \Theta_{\tilde{V}}(z)$ on a subregion $\Omega(S)$ of the unit disk which contains the origin. Then there exists a closely connected unitary colligation

$$(2.4.11) \qquad \tilde{U}_0 = \begin{pmatrix} \begin{pmatrix} \tilde{A}_0 & 0 \\ \tilde{B}_0 & \tilde{T} \end{pmatrix} & \begin{pmatrix} 0 \\ \tilde{F} \end{pmatrix} \\ (\tilde{C}_0 \quad \tilde{G}) & (\tilde{H}) \end{pmatrix} : \begin{pmatrix} \tilde{\mathfrak{H}}_0 \oplus \tilde{\mathfrak{H}} \\ \mathfrak{F} \end{pmatrix} \to \begin{pmatrix} \tilde{\mathfrak{H}}_0 \oplus \tilde{\mathfrak{H}} \\ \mathfrak{G} \end{pmatrix}$$

such that $\tilde{\mathfrak{H}}_0$ is a Hilbert space and $S(z) = \Theta_{\tilde{U}_0}(z)$ on $\Omega(S) \cap \Omega(\Theta_{\tilde{U}_0})$. If

$$(2.4.12) \qquad \tilde{U}_0' = \begin{pmatrix} \begin{pmatrix} \tilde{A}_0' & 0 \\ \tilde{B}_0' & \tilde{T} \end{pmatrix} & \begin{pmatrix} 0 \\ \tilde{F} \end{pmatrix} \\ (\tilde{C}_0' \quad \tilde{G}) & (\tilde{H}) \end{pmatrix} : \begin{pmatrix} \tilde{\mathfrak{H}}_0' \oplus \tilde{\mathfrak{H}} \\ \mathfrak{F} \end{pmatrix} \to \begin{pmatrix} \tilde{\mathfrak{H}}_0' \oplus \tilde{\mathfrak{H}} \\ \mathfrak{G} \end{pmatrix}$$

is a second closely connected unitary colligation such that $\tilde{\mathfrak{H}}_0'$ is a Hilbert space and $S(z) = \Theta_{\tilde{U}_0'}(z)$ on a neighborhood of the origin, there is an isomorphism $\tilde{W}_0$ from $\tilde{\mathfrak{H}}_0$ to $\tilde{\mathfrak{H}}_0'$ such that

$$(2.4.13) \qquad \begin{pmatrix} \tilde{W}_0 & 0 & 0 \\ 0 & 1_{\tilde{\mathfrak{H}}} & 0 \\ 0 & 0 & 1_{\mathfrak{G}} \end{pmatrix} \tilde{U}_0 = \tilde{U}_0' \begin{pmatrix} \tilde{W}_0 & 0 & 0 \\ 0 & 1_{\tilde{\mathfrak{H}}} & 0 \\ 0 & 0 & 1_{\mathfrak{F}} \end{pmatrix}.$$

In any choice of colligation (2.4.11), either $\tilde{\mathfrak{H}}_0$ is the zero space or it is infinite-dimensional and $\tilde{A}_0^*$ is a shift operator.

Again we are interested in conditions which imply that $\tilde{V}$ is unitary.

THEOREM 2.4.4. *Assume that $S(z) = \Theta_{\tilde{V}}(z)$ as in Theorem 2.4.3, and let the unitary operator (2.4.11) be chosen as in that result.*

(1) The operator $(\, \tilde{B}_0^* \quad \tilde{C}_0^* \,)$ is partially isometric with initial space $\ker \tilde{V}^*$ and final space $\operatorname{ran}(1 - \tilde{A}_0^*\tilde{A}_0)$. Furthermore,

$$(2.4.14) \qquad 1 - \tilde{V}\tilde{V}^* = \begin{pmatrix} \tilde{B}_0 \\ \tilde{C}_0 \end{pmatrix} (\, \tilde{B}_0^* \quad \tilde{C}_0^* \,)$$

and

$$(2.4.15) \qquad \begin{cases} \ker \tilde{V}^* = \operatorname{ran}\begin{pmatrix} \tilde{B}_0 \\ \tilde{C}_0 \end{pmatrix}, & \operatorname{ran} \tilde{V} = \ker (\, \tilde{B}_0^* \quad \tilde{C}_0^* \,), \\ \ker \tilde{B}_0^* = \operatorname{ran} \tilde{V} \cap \mathfrak{H}, & \ker \tilde{C}_0^* = \operatorname{ran} \tilde{V} \cap \mathfrak{G}. \end{cases}$$

(2) The identity

$$(2.4.16) \qquad \tilde{S}(z)\tilde{C}_0 = -F^*(1 - z\tilde{T}^*)^{-1}\tilde{B}_0$$

holds on a neighborhood of the origin, and $\ker \tilde{C}_0 \subseteq \ker \tilde{B}_0$.

(3) If $\mathfrak{N} = \{g : \tilde{S}(z)g \equiv 0\}$, then $\mathfrak{N} = \operatorname{ran} \tilde{C}_0|_{\ker \tilde{B}_0}$. Each of the following conditions implies that $\tilde{V}$ is unitary:

(i) $\tilde{C}_0 = 0$;
(ii) $\mathfrak{N} = \{0\}$ and $\tilde{B}_0 = 0$.
In this case $\tilde{\mathfrak{H}}_0 = \{0\}$ and $\tilde{U}_0 = \tilde{V}$.

Proofs of Theorems 2.4.3 and 2.4.4. These can be obtained by applying the previous results to $V = \tilde{V}^*$ or by parallel arguments. $\qquad \square$

2.5 Classes $\mathbf{S}_\kappa(\mathfrak{F}, \mathfrak{G})$

A. Definition and basic properties

We are now ready to unify the main realization theorems by showing that the hypotheses of Theorems 2.2.1, 2.2.2, and 2.2.3 are met for one and the same class of functions. More precisely, in Theorem 2.5.2 below it is shown that if $\mathfrak{F}$ and $\mathfrak{G}$ are Pontryagin spaces such that $\mathrm{ind}_-\,\mathfrak{F} = \mathrm{ind}_-\,\mathfrak{G}$, and if one of the four kernels (2.1.1)–(2.1.4) has κ negative squares, then all four have κ negative squares.

However, such a conclusion does not always hold when $\mathfrak{F}$ and $\mathfrak{G}$ are Kreĭn spaces (that is, without the condition $\mathrm{ind}_-\,\mathfrak{F} = \mathrm{ind}_-\,\mathfrak{G} < \infty$). See the example in §2.1 which shows that then it is possible to have $\mathrm{sq}_-\,K_S = 0$ and $\mathrm{sq}_-\,D_{\tilde{S}} = \mathrm{sq}_-\,D_S = \mathrm{sq}_-\,K_{\tilde{S}} = \infty$. We allow $\mathfrak{F}$ and $\mathfrak{G}$ to be Kreĭn spaces, but we exclude such examples in the following definition.

DEFINITION 2.5.1. *By the **generalized Schur class** $\mathbf{S}_\kappa(\mathfrak{F}, \mathfrak{G})$, where κ is any nonnegative integer and $\mathfrak{F}$ and $\mathfrak{G}$ are Kreĭn spaces, we mean the set of all $S(z)$ in $\mathbf{H}_0(\mathfrak{F}, \mathfrak{G})$ such that the kernels (2.1.1)–(2.1.4) all have κ negative squares. We write $\mathbf{S}_\kappa(\mathfrak{G})$ for this class when $\mathfrak{F} = \mathfrak{G}$.*

The elements of $\mathbf{S}_\kappa(\mathfrak{F}, \mathfrak{G})$ are called **generalized Schur functions**. Notice that $S(z) \equiv 0$ belongs to $\mathbf{S}_\kappa(\mathfrak{F}, \mathfrak{G})$ only when $\mathfrak{F}$ and $\mathfrak{G}$ are Hilbert spaces. If $S(z)$ belongs to $\mathbf{S}_\kappa(\mathfrak{F}, \mathfrak{G})$, then $\mathfrak{H}(S)$, $\mathfrak{H}(\tilde{S})$, $\mathfrak{D}(S)$, $\mathfrak{D}(\tilde{S})$ are all defined as Pontryagin spaces having negative index κ. If also $\mathrm{ind}_-\,\mathfrak{F} = \mathrm{ind}_-\,\mathfrak{G} < \infty$, which is the main case of interest here, then $S(z)$ has canonical coisometric, isometric, and unitary realizations as described in Theorems 2.2.1, 2.2.2, and 2.2.3. Our next result shows that in this case, it is sufficient to verify the condition on negative squares for only one of the kernels.

THEOREM 2.5.2. *Let $\mathfrak{F}$ and $\mathfrak{G}$ be Pontryagin spaces with $\mathrm{ind}_-\,\mathfrak{F} = \mathrm{ind}_-\,\mathfrak{G}$. If $S(z)$ is in $\mathbf{H}_0(\mathfrak{F}, \mathfrak{G})$ and any one of the kernels (2.1.1)–(2.1.4) has κ negative squares, then $S(z)$ belongs to $\mathbf{S}_\kappa(\mathfrak{F}, \mathfrak{G})$.*

A similar conclusion is obtained with different hypotheses in Theorem 4.3.8.

Proof. Suppose that $K_S(w, z)$ has κ negative squares. Apply Theorem 2.4.1 with (2.4.1) chosen as the canonical coisometric colligation in Theorem 2.2.1. Then $S(z) = \Theta_{U_0}(z)$ on $\Omega(S) \cap \Omega(\Theta_{U_0})$, where U_0 is a closely connected unitary colligation of the form (2.4.2) with $\mathfrak{H}_0$ a Hilbert space. By Theorem 2.1.2(3),

$$\mathrm{sq}_-\,K_S = \mathrm{ind}_-\,\mathfrak{H}(S) = \mathrm{ind}_-(\mathfrak{H}_0 \oplus \mathfrak{H}(S)) = \mathrm{sq}_-\,D_S.$$

The numbers of negative squares are computed on $\Omega(S) \cap \Omega(\Theta_{U_0})$, but they are the same on $\Omega(S)$ by Theorem 1.1.4. In particular, we have $\mathrm{sq}_-\,K_{\tilde{S}} < \infty$, since $\mathrm{sq}_-\, K_{\tilde{S}} \leq \mathrm{sq}_-\, D_S$. Repeating the argument with $S(z)$ replaced by $\tilde{S}(z)$, we obtain

$$\mathrm{sq}_-\, K_{\tilde{S}} = \mathrm{sq}_-\, D_{\tilde{S}}.$$

Since always $\mathrm{sq}_-\, D_S = \mathrm{sq}_-\, D_{\tilde{S}}$, all four kernels (2.1.1)–(2.1.4) have κ negative squares if $K_S(w, z)$ has κ negative squares.

In a similar way, all four kernels (2.1.1)–(2.1.4) have κ negative squares if $K_{\tilde{S}}(w, z)$ has κ negative squares. The finiteness of $\mathrm{sq}_-\, D_S = \mathrm{sq}_-\, D_{\tilde{S}}$ implies that of both $\mathrm{sq}_-\, K_S$ and $\mathrm{sq}_-\, K_{\tilde{S}}$, and thus the result follows. $\qquad\square$

Suppose that $S(z)$ belongs to $\mathbf{S}_\kappa(\mathfrak{F}, \mathfrak{G})$ for some Kreĭn spaces $\mathfrak{F}$ and $\mathfrak{G}$. Let $f \in \mathfrak{F}$ and $g \in \mathfrak{G}$ be fixed vectors. If $g = S(w_j)f$ for $\kappa + 1$ distinct points $w_1, \dots, w_{\kappa+1}$ in $\Omega(S)$, then $\langle g, g \rangle_\mathfrak{G} \leq \langle f, f \rangle_\mathfrak{F}$. To see this, argue by contradiction. If $\langle g, g \rangle_\mathfrak{G} > \langle f, f \rangle_\mathfrak{F}$, then the matrix

$$\left(\langle K_{\tilde{S}}(\bar{w}_j, \bar{w}_i)f, f \rangle_\mathfrak{F} \right)_{i,j=1}^{\kappa+1} = \left(\frac{\langle f, f \rangle_\mathfrak{F} - \langle g, g \rangle_\mathfrak{G}}{1 - \bar{w}_i w_j} \right)_{i,j=1}^{\kappa+1}$$

is nonpositive and invertible, contradicting the fact that $\mathrm{sq}_-\, K_{\tilde{S}} = \kappa$. When $\kappa = 0$, this simply says that the values of $S(z)$ are contraction operators.

There are other ways in which the values of a function $S(z)$ in $\mathbf{S}_\kappa(\mathfrak{F}, \mathfrak{G})$ behave like contraction operators. Recall that if $\mathfrak{H}$ and $\mathfrak{K}$ are Kreĭn spaces and $T \in \mathfrak{L}(\mathfrak{H}, \mathfrak{K})$ is a contraction, then $\ker T$ is a Hilbert subspace of $\mathfrak{H}$ (Dritschel and Rovnyak [1996], Theorem 2.7). The next result shows that the values of a generalized Schur function collectively have a similar property.

THEOREM 2.5.3. *Let $S(z)$ belong to $\mathbf{S}_\kappa(\mathfrak{F}, \mathfrak{G})$, where $\mathfrak{F}$ and $\mathfrak{G}$ are Pontryagin spaces such that $\mathrm{ind}_-\, \mathfrak{F} = \mathrm{ind}_-\, \mathfrak{G}$. Let $\mathfrak{M}$ be the closed subspace of all $f \in \mathfrak{F}$ such that $S(z)f \equiv 0$, and let $\mathfrak{N}$ be the closed subspace of all $g \in \mathfrak{G}$ such that $\tilde{S}(z)g \equiv 0$. Then $\mathfrak{M}$ is a Hilbert subspace of $\mathfrak{F}$, and $\mathfrak{N}$ is a Hilbert subspace of $\mathfrak{G}$. Write $\mathfrak{F} = \mathfrak{F}_0 \oplus \mathfrak{M}$ and $\mathfrak{G} = \mathfrak{G}_0 \oplus \mathfrak{N}$. Then*

(2.5.1) $$S(z) = Q_0 S(z) = S(z)P_0 = Q_0 S(z)P_0,$$

where P_0, Q_0 are the projection operators on $\mathfrak{F}, \mathfrak{G}$ with ranges $\mathfrak{F}_0, \mathfrak{G}_0$, respectively.

LEMMA 2.5.4. *Let $\mathfrak{H}, \mathfrak{K}$, and $\mathfrak{F}$ be Kreĭn spaces. Assume that $C \in \mathfrak{L}(\mathfrak{H}, \mathfrak{K})$, and let $D_C \in \mathfrak{L}(\mathfrak{D}_C, \mathfrak{K})$ be a defect operator for C^*. If*

$$V = (\, C \quad K \,) \in \mathfrak{L}(\mathfrak{H} \oplus \mathfrak{F}, \mathfrak{K})$$

satisfies $\mathrm{ind}_-\, (1 - V^*V) = \mathrm{ind}_-\, (1 - C^*C) < \infty$, then $K = D_C X$, where $X \in \mathfrak{L}(\mathfrak{F}, \mathfrak{D}_C)$ is a contraction.

Proof. This is a part of Dritschel and Rovnyak [1996], Theorem 3.2 (Generalized Form). $\qquad\square$

Proof of Theorem 2.5.3. By Theorem 2.2.1, $S(z) = \Theta_V(z)$ where

$$V = \begin{pmatrix} T & F \\ G & H \end{pmatrix} : \begin{pmatrix} \mathfrak{H}(S) \\ \mathfrak{F} \end{pmatrix} \to \begin{pmatrix} \mathfrak{H}(S) \\ \mathfrak{G} \end{pmatrix}$$

is the canonical coisometric colligation. Write $V = (\, C \quad K \,)$, where

$$C = \begin{pmatrix} T \\ G \end{pmatrix}, \qquad K = \begin{pmatrix} F \\ H \end{pmatrix}.$$

Since $VV^* = 1$, V is a bicontraction by Corollary 1.3.5. Thus

$$0 \le 1 - V^*V = \begin{pmatrix} 1 - C^*C & * \\ * & * \end{pmatrix},$$

and so $1 - C^*C \ge 0$. It follows that $\mathrm{ind}_- (1 - V^*V) = 0 = \mathrm{ind}_- (1 - C^*C)$. By Lemma 2.5.4, $K = D_C X$, where $D_C \in \mathfrak{L}(\mathfrak{D}_C, \mathfrak{H}(S) \oplus \mathfrak{G})$ is a defect operator for C^* and $X \in \mathfrak{L}(\mathfrak{F}, \mathfrak{D}_C)$ is a contraction. Now recalling that for any $f \in \mathfrak{F}$,

$$(Ff)(z) = \frac{S(z) - S(0)}{z} f,$$
$$Hf = S(0)f,$$

we see that the set $\mathfrak{M}$ of vectors $f \in \mathfrak{F}$ such that $S(z)f \equiv 0$ is the kernel of K, so $\mathfrak{M} = \ker K = \ker X$. Since the kernel of a contraction is a Hilbert space (Dritschel and Rovnyak [1996], Theorem 2.7), $\mathfrak{M}$ is a Hilbert subspace of $\mathfrak{F}$.

By construction, $S(z) = S(z)P_0$. Since $\tilde{S}(z)$ belongs to $\mathbf{S}_\kappa(\mathfrak{G}, \mathfrak{F})$ by Theorem 2.5.2, we can apply what has been shown to $\tilde{S}(z)$. It follows that $\mathfrak{N}$ is a Hilbert subspace of $\mathfrak{G}$ and $\tilde{S}(z) = \tilde{S}(z)Q_0$. Hence $S(z) = S(z)P_0 = Q_0 S(z)P_0$, and (2.5.1) follows. $\square$

When $\kappa = 0$ and $\mathfrak{F}$ and $\mathfrak{G}$ are Hilbert spaces, our definition of $\mathbf{S}_\kappa(\mathfrak{F}, \mathfrak{G})$ gives the same class as described in §2.1.

THEOREM 2.5.5. *Let $\mathfrak{F}$ and $\mathfrak{G}$ be Hilbert spaces, and suppose that $S(z)$ is a holomorphic function defined on a subregion of the unit disk containing the origin with values in $\mathfrak{L}(\mathfrak{F}, \mathfrak{G})$. Then $S(z)$ belongs to $\mathbf{S}_0(\mathfrak{F}, \mathfrak{G})$ if and only if it is the restriction of a holomorphic function $\hat{S}(z)$ on the unit disk whose values are contraction operators in $\mathfrak{L}(\mathfrak{F}, \mathfrak{G})$.*

Proof. If $S(z)$ belongs to $\mathbf{S}_0(\mathfrak{F}, \mathfrak{G})$, define $\hat{S}(z) = \Theta_V(z)$, $z \in \mathbf{D}$, where V is the canonical coisometric colligation associated with $S(z)$. This function agrees with $S(z)$ on $\Omega(S)$ by construction. It is holomorphic on the unit disk because, in the notation of Theorem 2.2.1, $\Theta_V(z) = H + zG(1 - zT)^{-1}F$, where T is a contraction operator on a Hilbert space. By Theorem 2.1.2(1), the kernel $K_{\hat{S}}(w, z)$ is nonnegative on the diagonal, hence the values of $\Theta_V(z)$ are contractions.

In the other direction, by Theorem 2.5.2, it is enough to show that the kernel $K_{\hat{S}}(w, z)$ is nonnegative if $\hat{S}(z)$ is holomorphic and has contractive values on the unit disk. This is well known and follows, for example, by methods in de Branges and Rovnyak [1966b] (Appendix, Lemma 5) and Sz.-Nagy and Foias [1970] (Proposition 8.1). $\qquad\square$

Theorem 4.3.7 generalizes Theorem 2.5.5 to the case in which $\mathfrak{F}$ and $\mathfrak{G}$ are Kreĭn spaces. The Kreĭn-Langer factorization in §4.2 characterizes the classes $\mathbf{S}_\kappa(\mathfrak{F}, \mathfrak{G})$ for arbitrary κ under the condition that $\mathfrak{F}$ and $\mathfrak{G}$ are Hilbert spaces.

B. Conformally invariant view

By Theorem 2.5.5, when $\mathfrak{F}$ and $\mathfrak{G}$ are Hilbert spaces, the generalized Schur class $\mathbf{S}_0(\mathfrak{F}, \mathfrak{G})$ can be interpreted as the set of holomorphic functions on $\mathbf{D}$ whose values are contraction operators in $\mathfrak{L}(\mathfrak{F}, \mathfrak{G})$. This class is conformally invariant: the composition of a function in the class with a conformal mapping of the unit disk onto itself again belongs to the class.

In general, the classes $\mathbf{S}_\kappa(\mathfrak{F}, \mathfrak{G})$ do not enjoy similar invariance properties. Functions in the classes are typically not holomorphic throughout $\mathbf{D}$. Indeed a choice of distinguished point in $\mathbf{D}$, namely 0, is implicit in the definition of $\mathbf{S}_\kappa(\mathfrak{F}, \mathfrak{G})$: functions are required to be holomorphic at 0, and even the algebraic forms of the three kernels $K_S(w, z), K_{\tilde{S}}(w, z), D_S(w, z)$ depend on the choice of this special point in Definition 2.5.1. However, there is an invariant formulation of the theory. It requires that the definitions be modified in such a way that 0 is replaced by a fixed number $\alpha \in \mathbf{D}$, which we call the **base point**.

First we define a notion of reflection relative to α. If $\alpha = 0$, this is simply complex conjugation. If $\alpha \neq 0$, the operation is reflection with respect to the line through α and the origin (Figure 2.5.1). Thus if $w \in \mathbf{C}$, we put

$$\bar{w}^\alpha = \alpha \bar{w}/\bar{\alpha}.$$

If $F(z)$ is an operator-valued function, let

$$\tilde{F}^\alpha(z) = F(\bar{z}^\alpha)^*$$

for all z such that the right side is defined. We call $\bar{w}^\alpha$ the α-**reflection** of w, and $\tilde{F}^\alpha(z)$ the α-**reflection** of $F(z)$. The case $\alpha = 0$ is included in these definitions by setting

$$\bar{w}^0 = \bar{w},$$

$$\tilde{F}^0(z) = \tilde{F}(z).$$

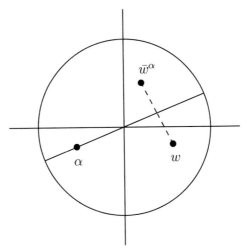

Figure 2.5.1

Let $\mathbf{H}_\alpha(\mathfrak{F}, \mathfrak{G})$ be the set of holomorphic functions $S(z)$ which are defined on a subregion $\Omega(S)$ of the unit disk containing α with values in $\mathfrak{L}(\mathfrak{F}, \mathfrak{G})$ for some Kreĭn spaces $\mathfrak{F}$ and $\mathfrak{G}$. For any $S(z)$ in $\mathbf{H}_\alpha(\mathfrak{F}, \mathfrak{G})$, the kernels

$$K_S^\alpha(w, z) = \frac{(1 - \bar{\alpha}z)(1 - \alpha\bar{w})}{1 - |\alpha|^2} \frac{1 - S(z)S(w)^*}{1 - z\bar{w}},$$

$$K_{\tilde{S}^\alpha}^\alpha(w, z) = \frac{(1 - \bar{\alpha}z)(1 - \alpha\bar{w})}{1 - |\alpha|^2} \frac{1 - \tilde{S}^\alpha(z)\tilde{S}^\alpha(w)^*}{1 - z\bar{w}},$$

$$D_S^\alpha(w, z) = \frac{(1 - \bar{\alpha}z)(1 - \alpha\bar{w})}{1 - |\alpha|^2} \begin{pmatrix} \dfrac{1 - S(z)S(w)^*}{1 - z\bar{w}} & \dfrac{\alpha}{|\alpha|} \dfrac{S(z) - \tilde{S}^\alpha(w)^*}{z - \bar{w}^\alpha} \\[3mm] \dfrac{\alpha}{|\alpha|} \dfrac{\tilde{S}^\alpha(z) - S(w)^*}{z - \bar{w}^\alpha} & \dfrac{1 - \tilde{S}^\alpha(z)\tilde{S}^\alpha(w)^*}{1 - z\bar{w}} \end{pmatrix},$$

are defined for all w and z in a subregion of the unit disk containing α. In the case $\alpha = 0$, the factor $\alpha/|\alpha|$ is interpreted as 1, and the three kernels reduce to those previously defined in (2.1.1)–(2.1.3):

$$K_S^0(w, z) = K_S(w, z),$$
$$K_{\tilde{S}^0}^0(w, z) = K_{\tilde{S}}(w, z),$$
$$D_S^0(w, z) = D_S(w, z).$$

We use these kernels to extend Definition 2.5.1.

CHAPTER 3

THE STATE SPACES

The spaces $\mathfrak{H}(S)$, $\mathfrak{H}(\tilde{S})$, $\mathfrak{D}(S)$ defined in Chapter 2 have many special properties when $S(z)$ belongs to $\mathbf{S}_\kappa(\mathfrak{F}, \mathfrak{G})$. Invariance under the difference-quotient transformation and an inequality characterize spaces of the form $\mathfrak{H}(S)$ for such functions (§3.1). Various conditions are derived for the canonical coisometric and isometric colligations to be unitary (§3.2, §3.3). Natural mappings relate the three spaces $\mathfrak{H}(S)$, $\mathfrak{H}(\tilde{S})$, $\mathfrak{D}(S)$ and lead to an operator range characterization of $\mathfrak{D}(S)$ in §3.4. A number of examples and applications are discussed in §3.5, including rational functions with unitary values on the unit circle.

3.1 Invariance under difference quotients

Which Pontryagin spaces are of the form $\mathfrak{H}(S)$? If no further constraint is imposed, the answer is surprising.

THEOREM 3.1.1. *Every reproducing kernel Pontryagin space of holomorphic functions on a subregion of the unit disk containing the origin, with values in a Kreĭn space $\mathfrak{G}$, is isometrically equal to a space $\mathfrak{H}(S)$ for some Kreĭn space $\mathfrak{F}$ and function $S(z)$ in $\mathbf{H}_0(\mathfrak{F}, \mathfrak{G})$.*

If the given reproducing kernel Pontryagin space and the space $\mathfrak{G}$ are finite-dimensional, so is the Kreĭn space $\mathfrak{F}$ which is constructed in the proof.

Proof. Let the given space be $\mathfrak{P}$, and assume that the functions in the space have domain Ω_0. If $E(w)$ is evaluation at a point $w \in \Omega_0$, the reproducing kernel for $\mathfrak{P}$ is given by $K(w, z) = E(z)E(w)^*$, $w, z \in \Omega_0$, by Theorem 1.1.2. Let $-\mathfrak{P}$ be the antispace of $\mathfrak{P}$, and define

$$S(z) = (\, 1_\mathfrak{G} \quad E(z) \quad zE(z) \,)$$

as an operator on $\mathfrak{F} = \mathfrak{G} \oplus (-\mathfrak{P}) \oplus \mathfrak{P}$ to $\mathfrak{G}$ for each $z \in \Omega_0$. Then

$$S(w)^* = \begin{pmatrix} 1_{\mathfrak{G}} \\ -E(w)^* \\ \bar{w}E(w)^* \end{pmatrix},$$

since the adjoint of $E(w)$ viewed as an operator from $-\mathfrak{P}$ to $\mathfrak{G}$ is $-E(w)^*$ (here $E(w)^*$ is the adjoint of $E(w)$ viewed as an operator from $\mathfrak{P}$ to $\mathfrak{G}$). Thus

$$K(w, z) = \frac{1_{\mathfrak{G}} - S(z)S(w)^*}{1 - z\bar{w}} = K_S(w, z), \qquad w, z \in \Omega_0.$$

In particular, $K_S(w, z)$ has finitely many squares. By Theorem 1.1.3, $\mathfrak{P}$ is equal isometrically to $\mathfrak{H}(S)$ since the reproducing kernels for the two spaces coincide. $\square$

Theorem 3.1.1 and the examples below are of a negative nature in that they show that with no further restriction, little can be proved.

EXAMPLE 1. *Let $\mathfrak{P}$ be the span of a holomorphic complex-valued function $h_0(z) \not\equiv 0$ defined on a subregion of the unit disk containing the origin, viewed as a Hilbert space in the norm which makes $h_0(z)$ a unit vector. By the construction in Theorem 3.1.1, $\mathfrak{P}$ is a space $\mathfrak{H}(S)$ for a function $S(z)$ in $\mathbf{H}_0(\mathfrak{F}, \mathfrak{G})$, namely*

$$S(z) = (\,1 \quad h_0(z) \quad zh_0(z)\,)$$

with $\mathfrak{F} = \mathbf{C} \oplus (-\mathbf{C}) \oplus \mathbf{C}$ and $\mathfrak{G} = \mathbf{C}$, where $-\mathbf{C}$ is the antispace of $\mathbf{C}$.

Since the properties of functions in the example are those of an arbitrary holomorphic function, nothing can be concluded in general about such matters as growth and holomorphic extensions into larger regions.

EXAMPLE 2. *A space $\mathfrak{P}$ can be of the form $\mathfrak{H}(S)$ in essentially different ways. For example, start with a space $\mathfrak{P} = \mathfrak{H}(S_1)$ for some $S_1(z)$ in $\mathbf{H}_0(\mathfrak{F}_1, \mathfrak{G})$ where $\mathfrak{F}_1$ and $\mathfrak{G}$ are Hilbert spaces, $\mathfrak{P} \neq \{0\}$. The construction in Theorem 3.1.1 gives a new representation $\mathfrak{P} = \mathfrak{H}(S_2)$ for $S_2(z)$ in $\mathbf{H}_0(\mathfrak{F}_2, \mathfrak{G})$ where*

$$\mathfrak{F}_2 = \mathfrak{G} \oplus (-\mathfrak{P}) \oplus \mathfrak{P}$$

is not a Hilbert space. Therefore it is impossible that the functions $S_1(z)$ and $S_2(z)$ are related by

$$S_1(z) \equiv S_2(z)W,$$

where W is an isomorphism from $\mathfrak{F}_1$ onto $\mathfrak{F}_2$.

Example 1 also shows that a space $\mathfrak{H}(S)$ is not necessarily invariant under the mapping $h(z)$ into $[h(z) - h(0)]/z$ as in Theorem 2.2.1. We call this mapping the **difference-quotient transformation**. With no further restriction, spaces $\mathfrak{H}(S)$ which are invariant under the difference-quotient transformation yield, up to isomorphism, every continuous operator on a Pontryagin space.

EXAMPLE 3. *Let $R \in \mathfrak{L}(\mathfrak{K})$ for some Pontryagin space $\mathfrak{K}$. Then there is a function $S(z)$ in $\mathbf{H}_0(\mathfrak{F}, \mathfrak{G})$ for some Kreĭn space $\mathfrak{F}$ and Pontryagin space $\mathfrak{G}$ such that a space $\mathfrak{H}(S)$ exists,*

$$T : h(z) \rightarrow [h(z) - h(0)]/z$$

is an everywhere defined transformation of $\mathfrak{H}(S)$ into itself, and R is unitarily equivalent to T. If $\mathfrak{K}$ is finite-dimensional, then $\mathfrak{F}$ and $\mathfrak{G}$ can also be chosen to be finite-dimensional.

In fact, let $\mathfrak{G} = \mathfrak{K}$, and let $\mathfrak{P}$ be the Pontryagin space of all holomorphic $\mathfrak{G}$-valued functions of the form $h(z) = (1 - zR)^{-1}k$, $k \in \mathfrak{K}$, in the inner product such that the mapping W which takes k to $h(z)$ is an isomorphism. The domain of functions in $\mathfrak{P}$ is taken to be a suitable neighborhood of the origin. It is easy to see that $\mathfrak{P}$ has the reproducing kernel

$$K_{\mathfrak{P}}(w, z) = (1 - zR)^{-1}(1 - \bar{w}R^*)^{-1}.$$

By Theorem 3.1.1, $\mathfrak{P}$ is isometrically equal to a space $\mathfrak{H}(S)$ for some Kreĭn space $\mathfrak{F}$ and function $S(z)$ in $\mathbf{H}_0(\mathfrak{F}, \mathfrak{G})$. When $\mathfrak{K}$ is finite-dimensional, the space $\mathfrak{F}$ constructed in the proof of Theorem 3.1.1 is finite-dimensional. In the general case, the difference-quotient transformation T is everywhere defined in $\mathfrak{H}(S)$ and $R = W^{-1}TW$.

Of greater interest are the spaces $\mathfrak{H}(S)$ with $S(z)$ in $\mathbf{S}_\kappa(\mathfrak{F}, \mathfrak{G})$ where $\mathfrak{F}$ is a Pontryagin space satisfying $\mathrm{ind}_- \mathfrak{F} = \mathrm{ind}_- \mathfrak{G}$. Such spaces occur as state spaces in canonical coisometric realizations as described in Theorem 2.2.1. They are invariant under the formation of difference quotients, and an inequality for the difference-quotient transformation turns out to characterize the spaces.

THEOREM 3.1.2. *Assume that $\mathfrak{P}$ is a reproducing kernel Pontryagin space of negative index κ whose elements are holomorphic functions on a subregion of the unit disk with values in a Pontryagin space $\mathfrak{G}$. Assume that $[h(z) - h(0)]/z$ belongs to the space whenever $h(z)$ is in the space, and that*

$$(3.1.1) \qquad \left\langle \frac{h(z) - h(0)}{z}, \frac{h(z) - h(0)}{z} \right\rangle_{\mathfrak{P}} \leq \langle h(z), h(z) \rangle_{\mathfrak{P}} - \langle h(0), h(0) \rangle_{\mathfrak{G}}.$$

Then $\mathfrak{P}$ is of the form $\mathfrak{H}(S)$ for some $S(z)$ in $\mathbf{S}_\kappa(\mathfrak{F}, \mathfrak{G})$ where $\mathfrak{F}$ is a Pontryagin space satisfying $\mathrm{ind}_- \mathfrak{F} = \mathrm{ind}_- \mathfrak{G}$. Moreover, in this representation the space $\mathfrak{F}$ and the function $S(z)$ can be chosen so that in addition $S(z)f \equiv 0$ only for $f = 0$. Conversely, (3.1.1) holds if $\mathfrak{P}$ is of the form $\mathfrak{H}(S)$ for some $S(z)$ in $\mathbf{S}_\kappa(\mathfrak{F}, \mathfrak{G})$ with $\mathrm{ind}_- \mathfrak{F} = \mathrm{ind}_- \mathfrak{G}$.

We call (3.1.1) the **difference-quotient inequality**. The statement that $\mathfrak{P}$ is of the form $\mathfrak{H}(S)$ allows for the possibility that the functions in $\mathfrak{P}$ and $\mathfrak{H}(S)$ are defined on different regions, but the spaces become isometrically equal when restricted to the same region.

Proof. Define operators $T \in \mathfrak{L}(\mathfrak{P})$ and $G \in \mathfrak{L}(\mathfrak{P}, \mathfrak{G})$ by

$$(Th)(z) = \frac{h(z) - h(0)}{z},$$
$$Gh = h(0),$$

on arbitrary elements of the space. Continuity of G is implied by the assumption that $\mathfrak{P}$ has a reproducing kernel. The operator T is continuous by the closed graph theorem. As in the proof of Theorem 2.2.1, we find that for w in a neighborhood of the origin,

$$((1 - wT)^{-1}h)(z) = \frac{zh(z) - wh(w)}{z - w}$$

for arbitrary elements of $\mathfrak{P}$. It follows that the evaluation mapping $E(w)$ on $\mathfrak{P}$ at w is given by $E(w) = G(1 - wT)^{-1}$. The reproducing kernel for $\mathfrak{P}$ thus has the form

(3.1.2) $K(w, z) = E(z)E(w)^* = G(1 - zT)^{-1}(1 - \bar{w}T^*)^{-1}G^*$

for w, z in a neighborhood of the origin by Theorem 1.1.2.

We use the data to build a closely outer connected coisometric colligation. Put

(3.1.3) $C = \begin{pmatrix} T \\ G \end{pmatrix} \in \mathfrak{L}(\mathfrak{P}, \mathfrak{P} \oplus \mathfrak{G}).$

By Theorem 1.3.4(1),

$$\text{ind}_- (1 - CC^*) + \text{ind}_- \mathfrak{P} = \text{ind}_- (1 - C^*C) + \text{ind}_- \mathfrak{P} + \text{ind}_- \mathfrak{G}.$$

Our hypothesis (3.1.1) in operator terms asserts that $T^*T + G^*G \le 1$, and hence $1 - C^*C \ge 0$. Therefore $\text{ind}_- (1 - C^*C) = 0$ and

$$\text{ind}_- (1 - CC^*) = \text{ind}_- \mathfrak{G}.$$

Take $\mathfrak{F} = \mathfrak{D}_C$ for any defect operator $D_C \in \mathfrak{L}(\mathfrak{D}_C, \mathfrak{P} \oplus \mathfrak{G})$ for C^*. By the definition of a defect operator in §1.3, D_C has zero kernel and $1 - CC^* = D_C D_C^*$. Define operators $F \in \mathfrak{L}(\mathfrak{F}, \mathfrak{P})$ and $H \in \mathfrak{L}(\mathfrak{F}, \mathfrak{G})$ by

$$D_C = \begin{pmatrix} F \\ H \end{pmatrix} \in \mathfrak{L}(\mathfrak{F}, \mathfrak{P} \oplus \mathfrak{G}).$$

Since $CC^* + D_C D_C^* = 1$,

$$V = \begin{pmatrix} T & F \\ G & H \end{pmatrix} : \begin{pmatrix} \mathfrak{P} \\ \mathfrak{F} \end{pmatrix} \to \begin{pmatrix} \mathfrak{P} \\ \mathfrak{G} \end{pmatrix}$$

is a coisometric colligation. The formula $E(w) = G(1 - wT)^{-1}$ obtained above may be used to show that the colligation is closely outer connected: a function $h \in \mathfrak{P}$ orthogonal to the range of $(1 - \bar{w}T^*)^{-1}G^*$ satisfies $h(w) = E(w)h = 0$ in a neighborhood of the origin and is therefore identically zero.

Let $S(z) = \Theta_V(z)$ on a subregion of the unit disk containing the origin. By construction, $\mathrm{ind}_-\, \mathfrak{F} = \mathrm{ind}_-\, \mathfrak{G}$. The function $S(z)$ belongs to $\mathbf{S}_\kappa(\mathfrak{F}, \mathfrak{G})$ by Theorems 2.1.2(1) and 2.5.2. By Theorem 2.1.2(1),

$$K_S(w, z) = G(1 - zT)^{-1}(1 - \bar{w}T^*)^{-1}G^* = K(w, z)$$

for w, z in a neighborhood of the origin. Hence $\mathfrak{P}$ and $\mathfrak{H}(S)$ have identical restrictions to a neighborhood of the origin.

If $f \in \mathfrak{F}$ and $S(z)f \equiv 0$, then

$$Hf + zG(1 - zT)^{-1}Ff \equiv 0,$$

so $Hf = 0$ and $Ff = 0$ because V is closely outer connected. Therefore $D_C f = 0$ and hence $f = 0$ because a defect operator is one-to-one.

It remains to show that (3.1.1) holds for any space $\mathfrak{H}(S)$ such that $S(z)$ belongs to $\mathbf{S}_\kappa(\mathfrak{F}, \mathfrak{G})$ where $\mathfrak{F}$ is a Pontryagin space satisfying $\mathrm{ind}_-\, \mathfrak{F} = \mathrm{ind}_-\, \mathfrak{G}$. Let

$$V = \begin{pmatrix} T & F \\ G & H \end{pmatrix} : \begin{pmatrix} \mathfrak{H}(S) \\ \mathfrak{F} \end{pmatrix} \to \begin{pmatrix} \mathfrak{H}(S) \\ \mathfrak{G} \end{pmatrix}$$

be the associated colligation in Theorem 2.2.1. Here T and G have the same meanings as above. Let $V = (\,C \quad K\,)$ with C given by (3.1.3). An argument in the proof of Theorem 2.5.3 shows that the operator in (3.1.3) satisfies

$$1 - T^*T - G^*G = 1 - C^*C \geq 0,$$

which is equivalent to (3.1.1). $\square$

THEOREM 3.1.3. *Suppose that two spaces $\mathfrak{H}(S_1)$ and $\mathfrak{H}(S_2)$ are equal isometrically. Assume that for each $j = 1, 2$, $S_j(z)$ is in $\mathbf{S}_\kappa(\mathfrak{F}_j, \mathfrak{G})$, $\mathfrak{F}_j$ and $\mathfrak{G}$ are Pontryagin spaces such that $\mathrm{ind}_-\, \mathfrak{F}_j = \mathrm{ind}_-\, \mathfrak{G}$, and $S_j(z)f_j \equiv 0$ only for $f_j = 0$. Then*

$$S_1(z) \equiv S_2(z)W$$

for some unitary operator $W \in \mathcal{L}(\mathfrak{F}_1, \mathfrak{F}_2)$.

Proof. Recall that $\mathfrak{H}(S_1)$ and $\mathfrak{H}(S_2)$ are equal isometrically if they coincide as Pontryagin spaces. By Theorem 1.1.2, the reproducing kernels for the spaces coincide, so

$$S_1(z)S_1(w)^* = S_2(z)S_2(w)^*, \qquad w, z \in \Omega_0,$$

where Ω_0 is the region on which the functions in $\mathfrak{H}(S_1)$ and $\mathfrak{H}(S_2)$ are defined. It follows that the linear relation $\mathbf{R}$ which is defined as the span of all pairs

$$\bigl(S_2(w)^*g, S_1(w)^*g\bigr), \qquad w \in \Omega_0, \ g \in \mathfrak{G},$$

is isometric. The hypotheses on $S_1(z)$ and $S_2(z)$ imply that $\mathbf{R}$ has dense domain and dense range. Hence $\mathbf{R}$ is the graph of a unitary operator $W^* \in \mathfrak{L}(\mathfrak{F}_2, \mathfrak{F}_1)$ by Theorem 1.4.2(3). The operator $W \in \mathfrak{L}(\mathfrak{F}_1, \mathfrak{F}_2)$ defined in this way has the required properties. $\qquad\square$

3.2 Spaces $\mathfrak{H}(S)$

We have seen in §3.1 that arbitrary spaces $\mathfrak{H}(S)$ are too general to have a meaningful theory. We now show that the subclass characterized in Theorem 3.1.2 is richly structured.

Throughout this section we fix $S(z)$ in $\mathbf{S}_\kappa(\mathfrak{F}, \mathfrak{G})$, where $\mathfrak{F}$ and $\mathfrak{G}$ are Pontryagin spaces such that $\mathrm{ind}_- \mathfrak{F} = \mathrm{ind}_- \mathfrak{G}$. Let

$$(3.2.1) \qquad V = \begin{pmatrix} T & F \\ G & H \end{pmatrix} : \begin{pmatrix} \mathfrak{H}(S) \\ \mathfrak{F} \end{pmatrix} \to \begin{pmatrix} \mathfrak{H}(S) \\ \mathfrak{G} \end{pmatrix}$$

be the canonical coisometric colligation associated with $S(z)$ in Theorem 2.2.1. Using Theorem 2.4.1, we also choose a Hilbert space $\mathfrak{H}_0$ and a closely connected unitary colligation

$$(3.2.2) \qquad U_0 = \left(\begin{pmatrix} A_0 & B_0 \\ 0 & T \\ (0 & G) \end{pmatrix} \begin{pmatrix} C_0 \\ F \\ (H) \end{pmatrix} \right) : \begin{pmatrix} \mathfrak{H}_0 \oplus \mathfrak{H}(S) \\ \mathfrak{F} \end{pmatrix} \to \begin{pmatrix} \mathfrak{H}_0 \oplus \mathfrak{H}(S) \\ \mathfrak{G} \end{pmatrix}$$

such that

$$(3.2.3) \qquad S(z) = \Theta_V(z) = \Theta_{U_0}(z), \qquad z \in \Omega(S).$$

Here we assume that the domain $\Omega(S)$ is suitably restricted so that both of the representations are valid.

The equality in (3.2.14) follows from Theorem 1.3.4(3). Use $U_0^* U_0 = 1$ to obtain

$$1 - T^* T = (G^* \quad B_0^*) \begin{pmatrix} G \\ B_0 \end{pmatrix}.$$

Hence for any $h_1, \ldots, h_n \in \mathfrak{H}(S)$,

$$\left(\langle (1 - T^* T) h_j, h_i \rangle_{\mathfrak{H}(S)} \right)_{i,j=1}^n = \left(\langle \begin{pmatrix} G \\ B_0 \end{pmatrix} h_j, \begin{pmatrix} G \\ B_0 \end{pmatrix} h_i \rangle_{\mathfrak{G} \oplus \mathfrak{H}_0} \right)_{i,j=1}^n.$$

Since $\mathfrak{H}_0$ is a Hilbert space, $\mathfrak{G} \oplus \mathfrak{H}_0$ is a Pontryagin space having the negative index ν. The number of negative eigenvalues of such a matrix cannot exceed ν by Lemma 1.1.1. This proves (3.2.14).

It remains to prove the last statement. It follows from (3.2.15) that for any $h_1, \ldots, h_n \in \mathfrak{H}(S)$,

$$\left(\langle (1 - T^* T) h_j, h_i \rangle_{\mathfrak{H}(S)} \right)_{i,j=1}^n = \left(\langle h_j(0), h_i(0) \rangle_{\mathfrak{G}} \right)_{i,j=1}^n$$
$$= \left(\langle G h_j, G h_i \rangle_{\mathfrak{G}} \right)_{i,j=1}^n.$$

The hypothesis that the closed span of all values of functions in $\mathfrak{H}(S)$ is all of $\mathfrak{G}$ implies that G has dense range. Hence by Lemma 1.1.1, the Gram matrix

$$\left(\langle G h_j, G h_i \rangle_{\mathfrak{G}} \right)_{i,j=1}^n$$

has ν negative eigenvalues for some choice of $h_1, \ldots, h_n \in \mathfrak{H}(S)$. Therefore equality holds in (3.2.14). $\square$

3.3 Spaces $\mathfrak{H}(\tilde{S})$

Parallel results hold for the state space $\mathfrak{H}(\tilde{S})$ associated with the canonical isometric realization, since it is defined by replacing $S(z)$ by its reflection $\tilde{S}(z)$. However, the difference-quotient transformation plays very different roles in the spaces $\mathfrak{H}(S)$ and $\mathfrak{H}(\tilde{S})$, in one case corresponding to the main transformation, and in the other to the adjoint of the main transformation of the colligation. The results for $\mathfrak{H}(\tilde{S})$ are, of course, very similar. Complete formulas are stated for reference purposes, but we omit or only give a hint for proofs in most cases.

Again fix $S(z)$ in $\mathbf{S}_\kappa(\mathfrak{F}, \mathfrak{G})$, where $\mathfrak{F}$ and $\mathfrak{G}$ are Pontryagin spaces such that $\text{ind}_- \mathfrak{F} = \text{ind}_- \mathfrak{G}$. Let

(3.3.1)
$$\tilde{V} = \begin{pmatrix} \tilde{T} & \tilde{F} \\ \tilde{G} & \tilde{H} \end{pmatrix} : \begin{pmatrix} \mathfrak{H}(\tilde{S}) \\ \mathfrak{F} \end{pmatrix} \to \begin{pmatrix} \mathfrak{H}(\tilde{S}) \\ \mathfrak{G} \end{pmatrix}$$

be the canonical isometric colligation associated with $S(z)$ in Theorem 2.2.2. By Theorem 2.4.3, there is a closely connected unitary colligation

$$(3.3.2) \qquad \tilde{U}_0 = \left(\begin{pmatrix} \tilde{A}_0 & 0 \\ \tilde{B}_0 & \tilde{T} \\ (\tilde{C}_0 & \tilde{G}) \end{pmatrix} \begin{pmatrix} 0 \\ \tilde{F} \\ (\tilde{H}) \end{pmatrix} \right) : \begin{pmatrix} \tilde{\mathfrak{H}}_0 \oplus \mathfrak{H}(\tilde{S}) \\ \mathfrak{F} \end{pmatrix} \rightarrow \begin{pmatrix} \tilde{\mathfrak{H}}_0 \oplus \mathfrak{H}(\tilde{S}) \\ \mathfrak{G} \end{pmatrix}$$

such that $\tilde{\mathfrak{H}}_0$ is a Hilbert space and

$$(3.3.3) \qquad\qquad S(z) = \Theta_{\tilde{V}}(z) = \Theta_{\tilde{U}_0}(z), \qquad z \in \Omega(S).$$

We restrict the domain $\Omega(\tilde{S})$ so that both representations are valid.

SUMMARY OF FORMULAS. *For each $k \in \mathfrak{H}(\tilde{S})$ there is a vector $\tilde{k}(0)$ in $\mathfrak{G}$ such that for all $f \in \mathfrak{F}$, $g \in \mathfrak{G}$, and $z \in \Omega(\tilde{S})$,*

$$(3.3.4) \qquad \begin{cases} (\tilde{T}k)(z) = zk(z) - \tilde{S}(z)\tilde{k}(0), \\ (\tilde{F}f)(z) = K_{\tilde{S}}(0, z)f, \\ \quad \tilde{G}k = \tilde{k}(0), \\ \quad \tilde{H}f = S(0)f, \end{cases}$$

and

$$(3.3.5) \qquad \begin{cases} (\tilde{T}^*k)(z) = \dfrac{k(z) - k(0)}{z}, \\ \quad \tilde{F}^*k = k(0), \\ (\tilde{G}^*g)(z) = \dfrac{\tilde{S}(z) - \tilde{S}(0)}{z} g, \\ \quad \tilde{H}^*g = \tilde{S}(0)g. \end{cases}$$

More generally, for each $k \in \mathfrak{H}(\tilde{S})$ and $w \in \Omega(\tilde{S})$ there is a vector $\tilde{k}(\bar{w})$ in $\mathfrak{G}$ such that for all $f \in \mathfrak{F}$, $g \in \mathfrak{G}$, and $z \in \Omega(\tilde{S})$,

$$(3.3.6) \qquad \begin{cases} ((1 - \bar{w}\tilde{T})^{-1}\tilde{T}k)(z) = \dfrac{zk(z) - \tilde{S}(z)\tilde{k}(\bar{w})}{1 - z\bar{w}}, \\ ((1 - \bar{w}\tilde{T})^{-1}\tilde{F}f)(z) = K_{\tilde{S}}(w, z)f, \\ \quad \tilde{G}(1 - \bar{w}\tilde{T})^{-1}k = \tilde{k}(\bar{w}), \end{cases}$$

and

$$(3.3.7) \quad \begin{cases} (\tilde{T}^*(1 - w\tilde{T}^*)^{-1}k)(z) = \dfrac{k(z) - k(w)}{z - w}, \\[2mm] \tilde{F}^*(1 - w\tilde{T}^*)^{-1}k = k(w), \\[2mm] ((1 - w\tilde{T}^*)^{-1}\tilde{G}^*g)(z) = \dfrac{\tilde{S}(z) - \tilde{S}(w)}{z - w}\, g. \end{cases}$$

In particular, evaluation at w on $\mathfrak{H}(\tilde{S})$ is given by $E(w) = \tilde{F}^*(1 - w\tilde{T}^*)^{-1}$. The vector $\tilde{k}(\bar{w})$ is the unique element of $\mathfrak{G}$ such that

$$(3.3.8) \qquad \left\langle \tilde{k}(\bar{w}), v \right\rangle_{\mathfrak{G}} = \left\langle k(z), \frac{\tilde{S}(z) - \tilde{S}(w)}{z - w}\, v \right\rangle_{\mathfrak{H}(\tilde{S})}$$

for all $v \in \mathfrak{G}$. Thus if $w \in \Omega(\tilde{S})$, $f \in \mathfrak{F}$, and

$$k(z) = K_{\tilde{S}}(w, z)f,$$

then

$$\tilde{k}(z) = \frac{S(z) - S(\bar{w})}{z - \bar{w}}\, f.$$

To derive these formulas, apply the corresponding results in §3.2 to $\tilde{V}^*$ (see the proof of Theorem 2.2.2).

A companion $\mathcal{B}(z)$ to the function $\mathcal{A}(z)$ in Theorem 3.2.1 appears in the analysis.

THEOREM 3.3.1. *Define a function with values in $\mathfrak{L}(\mathfrak{H}_0, \mathfrak{G})$ by*

$$(3.3.9) \qquad \mathcal{B}(z) = \tilde{C}_0 + z\tilde{G}(1 - z\tilde{T})^{-1}\tilde{B}_0.$$

Then

$$(3.3.10) \qquad K_S(w, z) = \frac{\mathcal{B}(z)\mathcal{B}(w)^*}{1 - z\bar{w}} + \tilde{G}(1 - z\tilde{T})^{-1}(1 - \bar{w}\tilde{T}^*)^{-1}\tilde{G}^*,$$

$$(3.3.11) \qquad \frac{1 - \tilde{A}_0^*\tilde{A}_0}{1 - z\bar{w}} = \frac{\tilde{B}(z)\tilde{B}(w)^*}{1 - z\bar{w}} + \tilde{B}_0^*(1 - z\tilde{T}^*)^{-1}(1 - \bar{w}\tilde{T})^{-1}\tilde{B}_0.$$

The colligation $\tilde{V}$ is unitary if and only if $\mathcal{B}(z) \equiv 0$.

If (3.3.2) is replaced by a unitary colligation

$$
\tilde{U}_0' = \left(\begin{pmatrix} \tilde{A}_0' & 0 \\ \tilde{B}_0' & \tilde{T} \\ (\tilde{C}_0' & \tilde{G}) \end{pmatrix} \begin{pmatrix} 0 \\ \tilde{F} \\ (\tilde{H}) \end{pmatrix} \right) : \begin{pmatrix} \tilde{\mathfrak{H}}_0' \oplus \mathfrak{H}(\tilde{S}) \\ \mathfrak{F} \end{pmatrix} \to \begin{pmatrix} \tilde{\mathfrak{H}}_0' \oplus \mathfrak{H}(\tilde{S}) \\ \mathfrak{G} \end{pmatrix}
$$

having the same properties and $\mathcal{B}'(z)$ is constructed from $\tilde{U}_0'$ in the same manner, then $\mathcal{B}'(z) = \mathcal{B}(z)\tilde{W}_0^*$ for some isomorphism $\tilde{W}_0$ from $\tilde{\mathfrak{H}}_0$ to $\tilde{\mathfrak{H}}_0'$ by Theorem 2.4.3.

Proof. Notice that we can write

$$
(S(z) \quad \mathcal{B}(z)) = (\tilde{H} \quad \tilde{C}_0) + z\tilde{G}(1 - z\tilde{T})^{-1} (\tilde{F} \quad \tilde{B}_0) = \Theta_{V_1}(z),
$$

where the colligation V_1 is given by

$$
V_1 = \begin{pmatrix} \tilde{T} & (\tilde{F} & \tilde{B}_0) \\ \tilde{G} & (\tilde{H} & \tilde{C}_0) \end{pmatrix} : \begin{pmatrix} \mathfrak{H}(\tilde{S}) \\ \begin{pmatrix} \mathfrak{F} \\ \tilde{\mathfrak{H}}_0 \end{pmatrix} \end{pmatrix} \to \begin{pmatrix} \mathfrak{H}(\tilde{S}) \\ \mathfrak{G} \end{pmatrix}.
$$

Since $\tilde{U}_0\tilde{U}_0^* = 1$, $V_1V_1^* = 1$. Hence by Theorem 2.1.2(1),

$$
\frac{1 - (S(z) \quad \mathcal{B}(z)) \begin{pmatrix} S(w)^* \\ \mathcal{B}(w)^* \end{pmatrix}}{1 - z\bar{w}} = \tilde{G}(1 - z\tilde{T})^{-1}(1 - \bar{w}\tilde{T}^*)^{-1}\tilde{G}^*,
$$

which is (3.3.10). In a similar way,

$$
(\tilde{A}_0^* \quad \tilde{\mathcal{B}}(z)) = (\tilde{A}_0^* \quad \tilde{C}_0^*) + z\tilde{B}_0^*(1 - z\tilde{T}^*)^{-1} (0 \quad \tilde{G}^*) = \Theta_{V_2}(z),
$$

where

$$
V_2 = \begin{pmatrix} \tilde{T}^* & (0 & \tilde{G}^*) \\ \tilde{B}_0^* & (\tilde{A}_0^* & \tilde{C}_0^*) \end{pmatrix} : \begin{pmatrix} \mathfrak{H}(\tilde{S}) \\ \begin{pmatrix} \tilde{\mathfrak{H}}_0 \\ \mathfrak{G} \end{pmatrix} \end{pmatrix} \to \begin{pmatrix} \mathfrak{H}(\tilde{S}) \\ \tilde{\mathfrak{H}}_0 \end{pmatrix}.
$$

Since $\tilde{U}_0^*\tilde{U}_0 = 1$, $V_2V_2^* = 1$. By Theorem 2.1.2(1),

$$
\frac{1 - (\tilde{A}_0^* \quad \tilde{\mathcal{B}}(z)) \begin{pmatrix} \tilde{A}_0 \\ \tilde{\mathcal{B}}(w)^* \end{pmatrix}}{1 - z\bar{w}} = \tilde{B}_0^*(1 - z\tilde{T}^*)^{-1}(1 - \bar{w}\tilde{T})^{-1}\tilde{B}_0,
$$

proving (3.3.11). The last assertion is proved as in the corresponding part of Theorem 3.2.1, but now using Theorem 2.4.4.   □

We shall also need analogs of Theorems 3.2.2–3.2.6. However, there is no need to derive these by parallel arguments as in the proof of Theorem 3.3.1, because they follow by the substitutions

$$S(z) \to \tilde{S}(z),$$

$$V \to \tilde{V}^*,$$

$$U_0 \to \tilde{U}_0^*,$$

$$\mathcal{A}(z) \to \mathcal{B}(z).$$

In fact, the proof of Theorem 2.2.2 shows that $\tilde{V}$ is the adjoint of the canonical coisometric colligation associated with $\tilde{S}(z)$. See also the proof of Theorem 2.4.3 for the connection between U_0 and $\tilde{U}_0^*$ and the definitions of $\mathcal{A}(z)$ and $\mathcal{B}(z)$ in (3.2.9) and (3.3.9). Theorems 3.3.2–3.3.6 all follow in this way from Theorems 3.2.2–3.2.6, and they are stated without proof.

THEOREM 3.3.2. (1) *For all* $h, k \in \mathfrak{H}(\tilde{S})$, $\alpha, \beta \in \Omega(\tilde{S})$,

$$\left\langle h(z) + \alpha \frac{h(z) - h(\alpha)}{z - \alpha}, k(z) + \beta \frac{k(z) - k(\beta)}{z - \beta} \right\rangle_{\mathfrak{H}(\tilde{S})}$$

$$- \left\langle \frac{h(z) - h(\alpha)}{z - \alpha}, \frac{k(z) - k(\beta)}{z - \beta} \right\rangle_{\mathfrak{H}(\tilde{S})}$$

$$= \langle h(\alpha), k(\beta) \rangle_{\mathfrak{F}} + \left\langle \tilde{B}_0^*(1 - \alpha \tilde{T}^*)^{-1} h, \tilde{B}_0^*(1 - \beta \tilde{T}^*)^{-1} k \right\rangle_{\mathfrak{H}_0}.$$

(2) *For all* $k \in \mathfrak{H}(\tilde{S})$, $\alpha, \beta \in \Omega(\tilde{S})$, *and* $g \in \mathfrak{G}$,

$$\beta \left\langle k(z), \frac{\tilde{S}(z) - \tilde{S}(\beta)}{z - \beta} g \right\rangle_{\mathfrak{H}(\tilde{S})} - (1 - \alpha \bar{\beta}) \left\langle \frac{k(z) - k(\alpha)}{z - \alpha}, \frac{\tilde{S}(z) - \tilde{S}(\beta)}{z - \beta} g \right\rangle_{\mathfrak{H}(\tilde{S})}$$

$$= \left\langle k(\alpha), \tilde{S}(\beta) g \right\rangle_{\mathfrak{F}} + \left\langle \mathcal{B}(\bar{\beta}) \tilde{B}_0^*(1 - \alpha \tilde{T}^*)^{-1} k, g \right\rangle_{\mathfrak{G}}.$$

(3) *For all* $\alpha, \beta \in \Omega(\tilde{S})$ *and* $g_1, g_2 \in \mathfrak{G}$,

$$\left\langle \frac{\tilde{S}(z) - \tilde{S}(\alpha)}{z - \alpha} g_1, \frac{\tilde{S}(z) - \tilde{S}(\beta)}{z - \beta} g_2 \right\rangle_{\mathfrak{H}(\tilde{S})}$$

$$= \langle K_S(\bar{\alpha}, \bar{\beta}) g_1, g_2 \rangle_{\mathfrak{G}} - \frac{1}{1 - \alpha \bar{\beta}} \left\langle \tilde{\mathcal{B}}(\alpha) g_1, \tilde{\mathcal{B}}(\beta) g_2 \right\rangle_{\mathfrak{H}_0}.$$

As in the parallel result in §3.2, condition (3) has a noteworthy consequence, namely,

$$(3.3.12) \quad \left\langle \sum_{j=1}^{n} \frac{\tilde{S}(z) - \tilde{S}(\alpha_j)}{z - \alpha_j} g_j, \sum_{i=1}^{n} \frac{\tilde{S}(z) - \tilde{S}(\alpha_i)}{z - \alpha_i} g_i \right\rangle_{\mathfrak{H}(\tilde{S})}$$

$$\leq \sum_{i,j=1}^{n} \langle K_S(\bar{\alpha}_j, \bar{\alpha}_i) g_j, g_i \rangle_{\mathfrak{G}}$$

for any $\alpha_1, \ldots, \alpha_n \in \Omega(\tilde{S})$ and $g_1, \ldots, g_n \in \mathfrak{G}$.

THEOREM 3.3.3. (1) *The following conditions are equivalent:*
(i) *if $\tilde{S}(z)g$ belongs to $\mathfrak{H}(\tilde{S})$ for some $g \in \mathfrak{G}$, then $\tilde{S}(z)g \equiv 0$;*
(ii) *$\tilde{B}_0 = 0$.*

(2) *These conditions are equivalent:*
(iii) *if $\tilde{S}(z)g$ belongs to $\mathfrak{H}(\tilde{S})$ for some $g \in \mathfrak{G}$, then $g = 0$;*
(iv) *$\tilde{V}$ is unitary.*

THEOREM 3.3.4. *The following are equivalent:*
(1) *For all $k \in \mathfrak{H}(\tilde{S})$,*

$$\left\langle \frac{k(z) - k(0)}{z}, \frac{k(z) - k(0)}{z} \right\rangle_{\mathfrak{H}(\tilde{S})} = \langle k(z), k(z) \rangle_{\mathfrak{H}(\tilde{S})} - \langle k(0), k(0) \rangle_{\mathfrak{F}}.$$

(2) *For all $h, k \in \mathfrak{H}(\tilde{S})$, $\alpha, \beta \in \Omega(\tilde{S})$,*

$$\left\langle h(z) + \alpha \frac{h(z) - h(\alpha)}{z - \alpha}, k(z) + \beta \frac{k(z) - k(\beta)}{z - \beta} \right\rangle_{\mathfrak{H}(\tilde{S})}$$

$$- \left\langle \frac{h(z) - h(\alpha)}{z - \alpha}, \frac{k(z) - k(\beta)}{z - \beta} \right\rangle_{\mathfrak{H}(\tilde{S})} = \langle h(\alpha), k(\beta) \rangle_{\mathfrak{F}}.$$

(3) *For all $k \in \mathfrak{H}(\tilde{S})$, $\alpha, \beta \in \Omega(\tilde{S})$, and $g \in \mathfrak{G}$,*

$$\beta \left\langle k(z), \frac{\tilde{S}(z) - \tilde{S}(\beta)}{z - \beta} g \right\rangle_{\mathfrak{H}(\tilde{S})}$$

$$- (1 - \alpha\bar{\beta}) \left\langle \frac{k(z) - k(\alpha)}{z - \alpha}, \frac{\tilde{S}(z) - \tilde{S}(\beta)}{z - \beta} g \right\rangle_{\mathfrak{H}(\tilde{S})} = \left\langle k(\alpha), \tilde{S}(\beta)g \right\rangle_{\mathfrak{F}}.$$

(4) *If $\tilde{S}(z)g$ belongs to $\mathfrak{H}(\tilde{S})$ for some $g \in \mathfrak{G}$, then $\tilde{S}(z)g \equiv 0$.*

THEOREM 3.3.5. *The following are equivalent:*

(1) *For all $k \in \mathfrak{H}(\tilde{S})$,*

$$\left\langle \frac{k(z) - k(0)}{z}, \frac{k(z) - k(0)}{z} \right\rangle_{\mathfrak{H}(\tilde{S})} = \langle k(z), k(z) \rangle_{\mathfrak{H}(\tilde{S})} - \langle k(0), k(0) \rangle_{\mathfrak{F}},$$

and the only $g \in \mathfrak{G}$ such that $\tilde{S}(z)g \equiv 0$ is $g = 0$.

(2) *For all $k \in \mathfrak{H}(\tilde{S})$ and $g \in \mathfrak{G}$,*

$$\left\langle \frac{k(z) - k(0)}{z}, \frac{\tilde{S}(z) - \tilde{S}(0)}{z} g \right\rangle_{\mathfrak{H}(\tilde{S})} = - \left\langle k(0), \tilde{S}(0)g \right\rangle_{\mathfrak{F}},$$

and the only $g \in \mathfrak{G}$ such that $\tilde{S}(z)g \equiv 0$ is $g = 0$.

(3) *For all $\alpha, \beta \in \Omega(\tilde{S})$ and $g_1, g_2 \in \mathfrak{G}$,*

$$\left\langle \frac{\tilde{S}(z) - \tilde{S}(\alpha)}{z - \alpha} g_1, \frac{\tilde{S}(z) - \tilde{S}(\beta)}{z - \beta} g_2 \right\rangle_{\mathfrak{H}(\tilde{S})} = \langle K_S(\bar{\alpha}, \bar{\beta}) g_1, g_2 \rangle_{\mathfrak{G}}.$$

(4) *The identity in (3) holds for $\alpha = \beta = 0$ and all $g_1, g_2 \in \mathfrak{G}$.*

(5) *The colligation $\tilde{V}$ is unitary.*

If the equivalent conditions hold, then $\tilde{V}$ is a Julia colligation if and only if the closed span of all values of functions in $\mathfrak{H}(\tilde{S})$ is all of $\mathfrak{F}$.

For any $k \in \mathfrak{H}(\tilde{S})$ define $\tilde{k}$ by (3.3.8).

THEOREM 3.3.6. *For any $k \in \mathfrak{H}(\tilde{S})$,*

(3.3.13)
$$\begin{cases} ((1 - \tilde{T}\tilde{T}^*)k)(z) = k(0) + \tilde{S}(z)(\tilde{T}^*k)^{\sim}(0), \\ ((1 - \tilde{T}^*\tilde{T})k)(z) = \dfrac{\tilde{S}(z) - \tilde{S}(0)}{z} \tilde{k}(0), \end{cases}$$

and

(3.3.14)
$$\mathrm{ind}_-\,(1 - \tilde{T}\tilde{T}^*) = \mathrm{ind}_-\,(1 - \tilde{T}^*\tilde{T}) \le \nu,$$

where $\nu = \mathrm{ind}_-\, \mathfrak{F} = \mathrm{ind}_-\, \mathfrak{G}$. When $\tilde{V}$ is unitary or, more generally, the equivalent conditions of Theorem 3.3.4 hold, then also for all $k \in \mathfrak{H}(\tilde{S})$,

(3.3.15)
$$((1 - \tilde{T}\tilde{T}^*)k)(z) = K_{\tilde{S}}(0, z)k(0).$$

If additionally the closed span of all values of functions in $\mathfrak{H}(\tilde{S})$ is all of $\mathfrak{F}$, then equality holds in (3.3.14).

3.4 Spaces $\mathfrak{D}(S)$

As in §3.2 and §3.3, fix $S(z)$ in $\mathbf{S}_\kappa(\mathfrak{F}, \mathfrak{G})$, where $\mathfrak{F}$ and $\mathfrak{G}$ are Pontryagin spaces such that $\mathrm{ind}_-\, \mathfrak{F} = \mathrm{ind}_-\, \mathfrak{G}$. Let

$$U = \begin{pmatrix} A & B \\ C & D \end{pmatrix} : \begin{pmatrix} \mathfrak{D}(S) \\ \mathfrak{F} \end{pmatrix} \to \begin{pmatrix} \mathfrak{D}(S) \\ \mathfrak{G} \end{pmatrix}$$

be the canonical unitary colligation associated with $S(z)$ in Theorem 2.3.1. In addition to the realizations V in (3.2.1)–(3.2.3) and $\tilde{V}$ in (3.3.1)–(3.3.3) we now also have

$$S(z) = \Theta_U(z).$$

The notation established in §3.2 and §3.3 is used here as well, and we assume that the domain $\Omega(S)$ is chosen so that $\Omega(S) = \Omega(\tilde{S})$ and all of the realizations are valid for $z \in \Omega(S)$.

The formulas for A, B, C, D and A^*, B^*, C^*, D^* which parallel (3.2.4)–(3.2.7) and (3.3.4)–(3.3.7) already appear in Theorem 2.3.1 and (2.3.4)–(2.3.5). In the case of spaces $\mathfrak{H}(S)$ and $\mathfrak{H}(\tilde{S})$, the theory divides according as the identity for difference quotients holds or not (see Theorems 3.2.4–3.2.5 and 3.3.4–3.3.5). An analogous identity holds with no additional hypotheses in spaces $\mathfrak{D}(S)$: for arbitrary elements of the space,

$$(3.4.1) \quad \left\langle \begin{pmatrix} [h(z) - h(0)]/z \\ zk(z) - \tilde{S}(z)h(0) \end{pmatrix}, \begin{pmatrix} [h(z) - h(0)]/z \\ zk(z) - \tilde{S}(z)h(0) \end{pmatrix} \right\rangle_{\mathfrak{D}(S)}$$

$$= \left\langle \begin{pmatrix} h(z) \\ k(z) \end{pmatrix}, \begin{pmatrix} h(z) \\ k(z) \end{pmatrix} \right\rangle_{\mathfrak{D}(S)} - \langle h(0), h(0) \rangle_{\mathfrak{G}}$$

and

$$(3.4.2) \quad \left\langle \begin{pmatrix} zh(z) - S(z)k(0) \\ [k(z) - k(0)]/z \end{pmatrix}, \begin{pmatrix} zh(z) - S(z)k(0) \\ [k(z) - k(0)]/z \end{pmatrix} \right\rangle_{\mathfrak{D}(S)}$$

$$= \left\langle \begin{pmatrix} h(z) \\ k(z) \end{pmatrix}, \begin{pmatrix} h(z) \\ k(z) \end{pmatrix} \right\rangle_{\mathfrak{D}(S)} - \langle k(0), k(0) \rangle_{\mathfrak{F}}.$$

These identities are immediate from the unitarity of the canonical coisometric colligation, which gives $A^*A + C^*C = 1$ and $AA^* + BB^* = 1$, and the formulas (2.3.1) and (2.3.2). Other identities from §3.2 and §3.3 have similar extensions for spaces $\mathfrak{D}(S)$ with no additional hypotheses, but we make no use of them. The phenomenon that spaces $\mathfrak{D}(S)$ have better properties than spaces $\mathfrak{H}(S)$ and $\mathfrak{H}(\tilde{S})$ appears also in behavior with respect to factorizations (see §4.1 B, Example 3).

The main point of interest here is to relate the state spaces $\mathfrak{H}(S)$, $\mathfrak{H}(\tilde{S})$, and $\mathfrak{D}(S)$ for the three realizations by means of natural operators that act among them.

THEOREM 3.4.1. *Let Λ be the mapping which takes any $h \in \mathfrak{H}(S)$ to the function $\tilde{h}$ defined by (3.2.8). Then Λ is a continuous bicontraction from $\mathfrak{H}(S)$ into $\mathfrak{H}(\tilde{S})$ such that*

$$(3.4.3) \qquad \Lambda : \sum_{j=1}^{n} K_S(w_j, z)g_j \to \sum_{j=1}^{n} \frac{\tilde{S}(z) - \tilde{S}(\bar{w}_j)}{z - \bar{w}_j} g_j$$

for any $w_1, \ldots, w_n \in \Omega(S)$ and $g_1, \ldots, g_n \in \mathfrak{G}$. The adjoint Λ^* is the mapping which takes any $k \in \mathfrak{H}(\tilde{S})$ to the function $\check{k}$ defined by (3.3.8). Hence Λ^* is a continuous bicontraction from $\mathfrak{H}(\tilde{S})$ into $\mathfrak{H}(S)$ such that

$$(3.4.4) \qquad \Lambda^* : \sum_{j=1}^{n} K_{\tilde{S}}(v_j, z)f_j \to \sum_{j=1}^{n} \frac{S(z) - S(\bar{v}_j)}{z - \bar{v}_j} f_j$$

for any $v_1, \ldots, v_n \in \Omega(S)$ and $f_1, \ldots, f_n \in \mathfrak{F}$. The mappings

$$(3.4.5) \qquad \Pi_1 : \begin{pmatrix} h \\ k \end{pmatrix} \to h, \qquad \Pi_2 : \begin{pmatrix} h \\ k \end{pmatrix} \to k$$

are coisometries from $\mathfrak{D}(S)$ onto $\mathfrak{H}(S)$ and $\mathfrak{H}(\tilde{S})$ with adjoints given by

$$(3.4.6) \qquad \Pi_1^* : h \to \begin{pmatrix} h \\ \Lambda h \end{pmatrix}, \qquad \Pi_2^* : k \to \begin{pmatrix} \Lambda^* k \\ k \end{pmatrix}.$$

for any $h \in \mathfrak{H}(S)$ and $k \in \mathfrak{H}(\tilde{S})$. In particular,

$$(3.4.7) \qquad \Lambda = \Pi_2 \Pi_1^*, \qquad \Lambda^* = \Pi_1 \Pi_2^*.$$

The kernel of Λ is zero if and only if V is closely inner connected, and the kernel of Λ^* is zero if and only if $\tilde{V}$ is closely outer connected.

Notice that the kernels of $\Lambda, \Lambda^*, \Pi_1, \Pi_2$ are all Hilbert spaces because Λ and Λ^* are contractions and $\mathfrak{H}(S)$, $\mathfrak{H}(\tilde{S})$, and $\mathfrak{D}(S)$ are Pontryagin spaces having the same negative index κ. The formulas (3.4.3) and (3.4.4) have the consequence that

$$D_S(w, \cdot) \begin{pmatrix} g \\ f \end{pmatrix} = \begin{pmatrix} 1 & \Lambda^* \\ \Lambda & 1 \end{pmatrix} \begin{pmatrix} K_S(w, \cdot) & 0 \\ 0 & K_{\tilde{S}}(w, \cdot) \end{pmatrix} \begin{pmatrix} g \\ f \end{pmatrix}$$

for all $f \in \mathfrak{F}$ and $g \in \mathfrak{G}$. The two-by-two operator matrix on the right side of this formula appears also in Theorem 3.4.3.

Proof. By (3.3.7),

$$\frac{\tilde{S}(z) - \tilde{S}(\bar{w})}{z - \bar{w}} g$$

belongs to $\mathfrak{H}(\tilde{S})$ whenever $w \in \Omega(S)$ and $g \in \mathfrak{G}$. Hence we may define a linear relation $\mathbf{R}$ from $\mathfrak{H}(S)$ to $\mathfrak{H}(\tilde{S})$ as the span of all pairs

$$(3.4.8) \qquad \left(K_S(w, z)g, \frac{\tilde{S}(z) - \tilde{S}(\bar{w})}{z - \bar{w}} g \right),$$

where $w \in \Omega(S)$ and $g \in \mathfrak{G}$. It is clear that $\mathbf{R}$ is densely defined. The identity

$$(3.4.9) \qquad \left\langle K_S(\bar{\alpha}, \bar{\beta})g_1, g_2 \right\rangle_{\mathfrak{G}} = \left\langle K_S(\bar{\alpha}, z)g_1, K_S(\bar{\beta}, z)g_2 \right\rangle_{\mathfrak{H}(S)}$$

holds for all $\alpha, \beta \in \Omega(S)$ and $g_1, g_2 \in \mathfrak{G}$. Thus by (3.3.12), $\mathbf{R}$ is a contraction. By Theorem 1.4.2(1) and Corollary 1.3.5, the closure of $\mathbf{R}$ is the graph of a continuous everywhere defined bicontraction Λ from $\mathfrak{H}(S)$ to $\mathfrak{H}(\tilde{S})$. The mapping h into $\tilde{h}$ is computed on special elements at the end of the Summary of Formulas in §3.2: when $h(z) = K_S(v, z)g$, the function defined by (3.2.8) is given by

$$\tilde{h}(z) = \frac{\tilde{S}(z) - \tilde{S}(\bar{v})}{z - \bar{v}} g,$$

and so $\Lambda h = \tilde{h}$ in this case. By linearity and continuity of function values, $\Lambda h = \tilde{h}$ for every $h \in \mathfrak{H}(S)$. The corresponding bicontraction $\tilde{\Lambda}$ from $\mathfrak{H}(\tilde{S})$ to $\mathfrak{H}(S)$ is constructed from the span of all pairs

$$(3.4.10) \qquad \left(K_{\tilde{S}}(v, z)f, \frac{S(z) - S(\bar{v})}{z - \bar{v}} f \right),$$

where $v \in \Omega(S)$ and $f \in \mathfrak{F}$. A routine calculation with the pairs (3.4.8) and (3.4.10) shows that $\tilde{\Lambda} = \Lambda^*$.

Next define a linear relation $\mathbf{R}_1$ from $\mathfrak{D}(S)$ to $\mathfrak{H}(S)$ as the span of pairs of the form

$$\left(D(w, z) \begin{pmatrix} g \\ f \end{pmatrix}, K_S(w, z)g + \frac{S(z) - S(\bar{w})}{z - \bar{w}} f \right),$$

where $w \in \Omega(S)$ and $f \in \mathfrak{F}$ and $g \in \mathfrak{G}$ are arbitrary vectors. The linear relation has dense domain, and the domain and range spaces are Pontryagin spaces having the same negative index κ. Consider numbers $\alpha_1, \ldots, \alpha_n \in \Omega(S)$ and vectors

$f_1, \ldots, f_n \in \mathfrak{F}$ and $g_1, \ldots, g_n \in \mathfrak{G}$. By (3.2.12),

$$\sum_{i,j=1}^{n} \left\langle K_S(\alpha_j, z)g_j + \frac{S(z) - S(\bar{\alpha}_j)}{z - \bar{\alpha}_j} f_j, K_S(\alpha_i, z)g_i + \frac{S(z) - S(\bar{\alpha}_i)}{z - \bar{\alpha}_i} f_i \right\rangle_{\mathfrak{H}(S)}$$

$$= \sum_{i,j=1}^{n} \left\{ \langle K_S(\alpha_j, \alpha_i)g_j, g_i \rangle_{\mathfrak{G}} + \left\langle \frac{S(\alpha_i) - S(\bar{\alpha}_j)}{\alpha_i - \bar{\alpha}_j} f_j, g_i \right\rangle_{\mathfrak{G}} \right.$$

$$+ \left\langle \frac{S(\alpha_j)^* - \tilde{S}(\alpha_i)}{\bar{\alpha}_j - \alpha_i} g_j, f_i \right\rangle_{\mathfrak{F}} + \left. \left\langle \frac{S(z) - S(\bar{\alpha}_j)}{z - \bar{\alpha}_j} f_j, \frac{S(z) - S(\bar{\alpha}_i)}{z - \bar{\alpha}_i} f_i \right\rangle_{\mathfrak{H}(S)} \right\}$$

$$\leq \sum_{i,j=1}^{n} \left\{ \langle K_S(\alpha_j, \alpha_i)g_j, g_i \rangle_{\mathfrak{G}} + \left\langle \frac{S(\alpha_i) - S(\bar{\alpha}_j)}{\alpha_i - \bar{\alpha}_j} f_j, g_i \right\rangle_{\mathfrak{G}} \right.$$

$$+ \left. \left\langle \frac{S(\alpha_j)^* - \tilde{S}(\alpha_i)}{\bar{\alpha}_j - \alpha_i} g_j, f_i \right\rangle_{\mathfrak{F}} + \langle K_{\tilde{S}}(\alpha_j, \alpha_i)f_j, f_i \rangle_{\mathfrak{F}} \right\}$$

$$= \sum_{i,j=1}^{n} \left\langle D(\alpha_j, z) \begin{pmatrix} g_j \\ f_j \end{pmatrix}, D(\alpha_i, z) \begin{pmatrix} g_i \\ f_i \end{pmatrix} \right\rangle_{\mathfrak{D}(S)},$$

which shows that $\mathbf{R}_1$ is a contraction. By Theorem 1.4.2(1) and Corollary 1.3.5, the closure of $\mathbf{R}_1$ is the graph of a continuous everywhere defined bicontraction on $\mathfrak{D}(S)$ into $\mathfrak{H}(S)$. By examining the dense set spanned by kernel functions, we see that this bicontraction is Π_1. Similarly, Π_2 acts as a bicontraction from $\mathfrak{D}(S)$ into $\mathfrak{H}(\tilde{S})$.

The formulas (3.4.6) are readily checked on kernel functions, and the general case follows by linearity and an approximation argument. By (3.4.5) and (3.4.6),

$$\Pi_1 \Pi_1^* = 1 \quad \text{and} \quad \Pi_2 \Pi_2^* = 1,$$

that is, Π_1 and Π_2 are coisometries. The same formulas yield (3.4.7).

By (3.4.4) and (3.2.6), $\overline{\operatorname{ran} \Lambda^*}$ is the same as the closed span of the ranges of the operators

$$(1 - wT)^{-1}F, \quad w \in \Omega(S),$$

and this is all of $\mathfrak{H}(S)$ if and only if V is closely inner connected. Thus Λ has zero kernel if and only if V is closely inner connected. The parallel assertion for Λ^* is obtained similarly. $\quad\square$

THEOREM 3.4.2. (A) *The following are equivalent:*

(1) Λ *is isometric;*
(2) Π_2 *is unitary;*
(3) $\tilde{V}$ *is unitary.*

When (1)–(3) *hold,* $\ker \Lambda^*$ *is a Hilbert subspace of* $\mathfrak{H}(\tilde{S})$ *equal to the set of functions* $k(z)$ *such that* $z^n k(z) \in \mathfrak{H}(\tilde{S})$ *for all* $n = 0, 1, 2, \ldots$. *Moreover,* $\ker \Lambda^*$ *is invariant under* $\tilde{T}$ *and* $\tilde{T}|_{\ker \Lambda^*}$ *is an isometry which acts like multiplication by* z.

(B) *These assertions are equivalent:*

(4) Λ^* *is isometric;*
(5) Π_1 *is unitary;*
(6) V *is unitary.*

When (4)–(6) *hold,* $\ker \Lambda$ *is a Hilbert subspace of* $\mathfrak{H}(S)$ *equal to the set of functions* $h(z)$ *such that* $z^n h(z) \in \mathfrak{H}(S)$ *for all* $n = 0, 1, 2, \ldots$. *Moreover,* $\ker \Lambda$ *is invariant under* T^* *and* $T^*|_{\ker \Lambda}$ *is an isometry which acts like multiplication by* z.

(C) *These assertions are equivalent:*

(7) Λ *is unitary;*
(8) Π_1 *and* Π_2 *are unitary;*
(9) V *and* $\tilde{V}$ *are unitary.*

Proof. We only prove (A) explicitly. The proof of (B) is similar, and (C) is obtained by combining (A) and (B).

The equivalence of (1) and (3) follows from (3.4.3) and the equivalence of the conditions in Theorem 3.3.5(3) and 3.3.5(5). By Theorem 3.4.1, $\Lambda = \Pi_2 \Pi_1^*$ and Π_1^* is isometric, hence (2) implies (1). We show that (3) implies (2). Assume that $\tilde{V}$ is unitary. To see that Π_2 is isometric, we need only check the action on kernel functions. This reduces to showing that

$$
\left\langle \frac{\tilde{S}(z) - \tilde{S}(\bar{\alpha})}{z - \bar{\alpha}} g_1 + K_{\tilde{S}}(\alpha, z) f_1, \frac{\tilde{S}(z) - \tilde{S}(\bar{\beta})}{z - \bar{\beta}} g_2 + K_{\tilde{S}}(\beta, z) f_2 \right\rangle_{\mathfrak{H}(\tilde{S})}
$$

$$
= \left\langle D_S(\alpha, z) \begin{pmatrix} g_1 \\ f_1 \end{pmatrix}, D_S(\beta, z) \begin{pmatrix} g_2 \\ f_2 \end{pmatrix} \right\rangle_{\mathfrak{D}(S)}
$$

for all $f_1, f_2 \in \mathfrak{F}$, $g_1, g_2 \in \mathfrak{G}$, and $\alpha, \beta \in \Omega(S)$. This follows on expanding and using the identity in Theorem 3.3.5(3).

Assume (1)–(3) hold. Then $\ker \Lambda^*$ is a Hilbert subspace of $\mathfrak{H}(\tilde{S})$ because Λ^* is a bicontraction (for example, see Dritschel and Rovnyak [1996], Theorem 2.7).

Suppose $z^n k(z) \in \mathfrak{H}(\tilde{S})$ for all $n \geq 0$. By (3.3.4),

$$(\tilde{T}k)(z) = zk(z) - \tilde{S}(z)\tilde{G}k.$$

Our hypotheses yield $\tilde{S}(z)\tilde{G}k \in \mathfrak{H}(\tilde{S})$. But by Theorem 3.3.3(2), $\tilde{S}(z)g \in \mathfrak{H}(\tilde{S})$ only for $g = 0$ since $\tilde{V}$ is unitary. Thus $(\tilde{T}k)(z) = zk(z)$ and $\tilde{G}k = 0$. Then

$$(\tilde{T}^2 k)(z) = z^2 k(z) - \tilde{S}(z)\tilde{G}\tilde{T}k,$$

and $(\tilde{T}^2 k)(z) = z^2 k(z)$ and $\tilde{G}\tilde{T}k = 0$. An inductive argument thus shows that $(\tilde{T}^{n+1}k)(z) = z^{n+1}k(z)$ and $\tilde{G}\tilde{T}^n k = 0$ for all $n \geq 0$. Hence $\tilde{G}(1 - w\tilde{T})^{-1}k = 0$. Then by (3.3.8) and (3.3.6),

$$\left\langle k(z), \frac{\tilde{S}(z) - \tilde{S}(\bar{w})}{z - \bar{w}} g \right\rangle_{\mathfrak{H}(\tilde{S})} = \left\langle \tilde{G}(1 - w\tilde{T})^{-1}k, g \right\rangle_{\mathfrak{G}} = 0$$

for all $w \in \Omega(S)$ and $g \in \mathfrak{G}$, and so $k \in (\operatorname{ran} \Lambda)^{\perp} = \ker \Lambda^*$. These steps are reversible, and thus

$$\ker \Lambda^* = \left\{ k(z) : z^n k(z) \in \mathfrak{H}(\tilde{S}), \; n \geq 0 \right\}.$$

The remaining assertions in (A) follow from the construction. We only remark that the fact that multiplication by z is an isometry on $\ker \Lambda^*$ follows from Theorem 3.3.5(1). $\qquad\square$

In view of the continuity of the mappings (3.4.5), $\mathfrak{D}(S)$ is contained continuously in $\mathfrak{H}(S) \oplus \mathfrak{H}(\tilde{S})$. This observation leads to a characterization of the space. Recall from Theorem 1.5.1 that if $\mathfrak{H}$ is a Pontryagin space and $P_1 \in \mathfrak{L}(\mathfrak{H})$ is a selfadjoint operator such that $\operatorname{ind}_- P_1 < \infty$, there is a unique Pontryagin space $\mathfrak{H}_1$ which is contained continuously in $\mathfrak{H}$ such that the range of P_1 is contained in $\mathfrak{H}_1$ as a dense subspace and

$$(3.4.11) \qquad \langle P_1 f, P_1 g \rangle_{\mathfrak{H}_1} = \langle P_1 f, g \rangle_{\mathfrak{H}}, \qquad f, g \in \mathfrak{H}.$$

The positive and negative indices of $\mathfrak{H}_1$ coincides with those of P_1.

THEOREM 3.4.3. *Let P_1 be the selfadjoint operator*

$$(3.4.12) \qquad P_1 = \begin{pmatrix} 1 & \Lambda^* \\ \Lambda & 1 \end{pmatrix}$$

on $\mathfrak{H}(S) \oplus \mathfrak{H}(\tilde{S})$. Then $\operatorname{ind}_- P_1 < \infty$ and the corresponding Pontryagin space $\mathfrak{H}_1$ provided by Theorem 1.5.1 is equal isometrically to $\mathfrak{D}(S)$. If Λ is unitary, $\frac{1}{2}P_1$ is a projection. In this case, $\mathfrak{D}(S)$ coincides with the range of $\frac{1}{2}P_1$ as a vector space with inner product given by

$$(3.4.13) \qquad \left\langle \begin{pmatrix} h_1 \\ k_1 \end{pmatrix}, \begin{pmatrix} h_2 \\ k_2 \end{pmatrix} \right\rangle_{\mathfrak{D}(S)} = 2\langle h_1, h_2 \rangle_{\mathfrak{H}(S)} + 2\langle k_1, k_2 \rangle_{\mathfrak{H}(\tilde{S})}$$

for all elements of $\mathfrak{D}(S)$.

Proof. The inclusion mapping from $\mathfrak{D}(S)$ into $\mathfrak{H}(S) \oplus \mathfrak{H}(\tilde{S})$ is continuous and given by

$$E_1 = \begin{pmatrix} \Pi_1 \\ \Pi_2 \end{pmatrix}.$$

Thus

$$E_1 E_1^* = \begin{pmatrix} \Pi_1 \\ \Pi_2 \end{pmatrix} \begin{pmatrix} \Pi_1^* & \Pi_2^* \end{pmatrix} = \begin{pmatrix} \Pi_1 \Pi_1^* & \Pi_1 \Pi_2^* \\ \Pi_2 \Pi_1^* & \Pi_2 \Pi_2^* \end{pmatrix} = \begin{pmatrix} 1 & \Lambda^* \\ \Lambda & 1 \end{pmatrix} = P_1.$$

It follows that $\text{ind}_- P_1 \leq \text{ind}_- \mathfrak{D}(S) < \infty$, and so $\mathfrak{H}_1 = \mathfrak{D}(S)$ isometrically by the uniqueness part of Theorem 1.5.1. Since Λ and Λ^* are bicontractions,

$$P_1^2 = \begin{pmatrix} 1 + \Lambda^* \Lambda & 2\Lambda^* \\ 2\Lambda & 1 + \Lambda \Lambda^* \end{pmatrix} \leq 2 \begin{pmatrix} 1 & \Lambda^* \\ \Lambda & 1 \end{pmatrix} = 2P_1,$$

with equality if Λ is unitary. Thus $\frac{1}{2} P_1$ is a projection on $\mathfrak{H}(S) \oplus \mathfrak{H}(\tilde{S})$ when Λ is unitary. Then using (3.4.11) applied with P_1 given by (3.4.12), we see that $\mathfrak{D}(S)$ coincides with $\text{ran}\, \frac{1}{2} P_1$ with inner product given by (3.4.13).   □

There is another way to view $\mathfrak{D}(S)$. To describe it, we assume that $\mathfrak{F} = \mathfrak{G}$, and so $S(z)$ belongs to $\mathbf{S}_\kappa(\mathfrak{G})$. It is always possible to reduce to this case by embedding $\mathfrak{F}$ and $\mathfrak{G}$ isometrically in a single Pontryagin space, which we can choose to have the same negative index as $\mathfrak{F}$ and $\mathfrak{G}$. Let

$$\Omega_+ = \Omega(S),$$
$$\Omega_- = \{z : 1/z \in \Omega_+\}.$$

For each $\begin{pmatrix} h \\ k \end{pmatrix} \in \mathfrak{D}(S)$, define a function F on $\Omega_+ \cup \Omega_-$ with values in $\mathfrak{G}$ by

$$(3.4.14) \qquad F(z) = \begin{cases} h(z), & z \in \Omega_+, \\ k(1/z), & z \in \Omega_-. \end{cases}$$

Let $\hat{\mathfrak{D}}(S)$ be the set of all functions F of this form, viewed as a Pontryagin space in the inner product such that the correspondence (3.4.14) is an isomorphism. The reproducing kernel for $\hat{\mathfrak{D}}(S)$ is easily computed and expressed in terms of the function

$$(3.4.15) \qquad \hat{S}(z) = \begin{cases} S(z), & z \in \Omega_+, \\ S(1/\bar{z})^*, & z \in \Omega_-. \end{cases}$$

It is given by

$$(3.4.16) \qquad \hat{K}_S(w, z) = \begin{cases} \dfrac{1 - \hat{S}(z)\hat{S}(w)^*}{1 - z\bar{w}}, & w \in \Omega_+, \ z \in \Omega_+, \\[2ex] \dfrac{\hat{S}(z) - \hat{S}(w)^*}{1/z - \bar{w}}, & w \in \Omega_+, \ z \in \Omega_-, \\[2ex] \dfrac{\hat{S}(z) - \hat{S}(w)^*}{z - 1/\bar{w}}, & w \in \Omega_-, \ z \in \Omega_+, \\[2ex] \dfrac{1 - \hat{S}(z)\hat{S}(w)^*}{1 - 1/(z\bar{w})}, & w \in \Omega_-, \ z \in \Omega_-. \end{cases}$$

We may thus alternatively think of $\mathfrak{D}(S)$ as the space of holomorphic $\mathfrak{G}$-valued functions on $\Omega_+ \cup \Omega_-$ with the reproducing kernel (3.4.16), where $\hat{S}(z)$ is defined by (3.4.15).

3.5 Examples and miscellaneous results

A. Rational unitary functions

This special case provides concrete examples of the classes $\mathbf{S}_\kappa(\mathfrak{F}, \mathfrak{G})$. It also allows us to compare the present approach to realization theory with the approach based on minimality. Examples show that, in general, the approach based on canonical coisometric, isometric, and unitary realizations is different from that based on minimality. However, in the case of **rational unitary functions**, that is, rational functions with unitary values on the unit circle, excluding poles, the two methods are essentially the same.

For simplicity, we take $\mathfrak{F}$ and $\mathfrak{G}$ to be finite-dimensional Kreĭn spaces. By a rational function we mean a finite sum of expressions which are polynomials in z or $1/(z - \lambda)$, $\lambda \in \mathbf{C}$, with coefficients in $\mathfrak{L}(\mathfrak{F}, \mathfrak{G})$. The term "pole" has its usual meaning for operator-valued holomorphic functions (we do not need any of the finer structure such as "order").

THEOREM 3.5.1. *Let $\mathfrak{F}$ and $\mathfrak{G}$ be finite-dimensional Kreĭn spaces such that* $\operatorname{ind}_- \mathfrak{F} = \operatorname{ind}_- \mathfrak{G}$. *The following are equivalent:*

(1) *$S(z) \in \mathbf{S}_\kappa(\mathfrak{F}, \mathfrak{G})$ for some nonnegative integer κ and $\dim \mathfrak{H}(S) < \infty$;*
(2) *$S(z)$ is a rational function with values in $\mathfrak{L}(\mathfrak{F}, \mathfrak{G})$ which is holomorphic at the origin and such that $S(z)S(z)^* = 1$ at all points of the circle $|z| = 1$ excluding poles.*

For an illustration of the possibilities, see Example 2 in the Appendix. Example 2 in §4.5 can also be adapted to fit the situation of Theorem 3.5.1.

Proof. $(1) \Rightarrow (2)$ If (1) holds, then by Theorem 2.2.1,

$$S(z) = \Theta_V(z) = H + zG(1 - zT)^{-1}F,$$

where

$$V = \begin{pmatrix} T & F \\ G & H \end{pmatrix} : \begin{pmatrix} \mathfrak{H}(S) \\ \mathfrak{F} \end{pmatrix} \to \begin{pmatrix} \mathfrak{H}(S) \\ \mathfrak{G} \end{pmatrix}$$

is a coisometric and closely outer connected colligation. Since $\mathfrak{H}(S)$ is finite-dimensional, $S(z)$ is rational by the form of the resolvent of an operator on a finite-dimensional space. By Theorem 2.1.2(1),

$$1 - S(z)S(w)^* = (1 - z\bar{w})G(1 - zT)^{-1}(1 - \bar{w}T^*)^{-1}G^*,$$

so $S(z)S(z)^* = 1$ on the unit circle except for poles, and hence (2) follows.

$(2) \Rightarrow (1)$ Assume (2). Writing $w = 1/\bar{v}$, we obtain

$$\begin{aligned} K_S(w, z) &= \frac{1 - S(z)S(w)^*}{1 - z\bar{w}} \\ &= \frac{S(z)S(z)^* - S(z)S(1/\bar{v})^*}{1 - z/v} \\ &= -\frac{S(z)}{1/v} \frac{\tilde{S}(1/z) - \tilde{S}(1/v)}{z - v} \end{aligned}$$

first for points z on the unit circle, and then by analyticity for all z excluding poles. By the definition of a rational function, the span of functions of the form

$$\frac{\tilde{S}(1/z) - \tilde{S}(1/v)}{z - v} g,$$

for g in $\mathfrak{G}$ and v in a neighborhood of infinity, is finite-dimensional. Therefore the span $\mathfrak{H}_0$ of functions $K_S(w, \cdot)g$, for w in a neighborhood of the origin and $g \in \mathfrak{G}$, is finite-dimensional. Introduce an inner product on $\mathfrak{H}_0$ as in the proof of Theorem 1.1.3. Applying Lemma 1.1.1' to the Gram matrices

$$\left(\langle K_S(w_j, \cdot)g_j, K_S(w_i, \cdot)g_i \rangle_{\mathfrak{H}_0} \right)_{i,j=1}^n = \left(\langle K_S(w_j, w_i)g_j, g_i \rangle_{\mathfrak{G}} \right)_{i,j=1}^n,$$

we see by the finite-dimensionality of $\mathfrak{H}_0$ that $K_S(w, z)$ has a finite number of negative squares, say $\mathrm{sq}_- K_S = \kappa$. Since $S(z)$ is holomorphic at the origin, $S(z) \in \mathbf{S}_\kappa(\mathfrak{F}, \mathfrak{G})$ by Theorem 2.5.2. By construction, $\mathfrak{H}_0$ and $\mathfrak{H}(S)$ coincide as vector spaces, and therefore $\dim \mathfrak{H}(S) < \infty$. $\square$

Realization theory for rational functions is often approached in a manner different from that presented in Chapter 2. Following our conventions in §1.2, we continue to assume that all underlying spaces are Kreĭn spaces. It should be noticed, however, that inner products play no role in some places, such as the definition of minimality of a realization.

DEFINITION 3.5.2. *Let $S(z)$ be a rational function which is holomorphic at the origin with values in $\mathfrak{L}(\mathfrak{F}, \mathfrak{G})$, where $\mathfrak{F}$ and $\mathfrak{G}$ are finite-dimensional Kreĭn spaces. A realization $S(z) = \Theta_V(z)$,*

$$(3.5.1) \qquad V = \begin{pmatrix} T & F \\ G & H \end{pmatrix} : \begin{pmatrix} \mathfrak{H} \\ \mathfrak{F} \end{pmatrix} \to \begin{pmatrix} \mathfrak{H} \\ \mathfrak{G} \end{pmatrix},$$

is called **minimal** *if $\mathfrak{H}$ is a finite-dimensional Kreĭn space and $\dim \mathfrak{H}$ is as small as possible for all such representations.*

The colligation (3.5.1) models a system

whose state in discrete time is described by vectors $h_0, h_1, h_2, \ldots$ in $\mathfrak{H}$. Let $f_0, f_1, f_2, \ldots$ and $g_0, g_1, g_2, \ldots$ belong to $\mathfrak{F}$ and $\mathfrak{G}$, and assume that $h_0 = 0$ and

$$(3.5.2) \qquad \begin{cases} h_{n+1} = Th_n + Ff_n, \\ \quad g_n = Gh_n + Hf_n, \end{cases}$$

for all $n \geq 0$. We call $\mathfrak{H}$ the **state space** and $\mathfrak{F}$ and $\mathfrak{G}$ the **input** and **output spaces** of the system. In terms of the formal power series $f(z) = \sum_{n=0}^{\infty} f_n z^n$, $g(z) = \sum_{n=0}^{\infty} g_n z^n$, $h(z) = \sum_{n=0}^{\infty} h_n z^n$, the relations (3.5.2) yield

$$h(z)/z = Th(z) + Ff(z),$$
$$g(z) = Gh(z) + Hf(z).$$

Eliminating $h(z)$, we obtain

$$g(z) = \left[H + zG(1 - zT)^{-1}F \right] f(z) = \Theta_V(z)f(z).$$

The characteristic function $\Theta_V(z)$ of the colligation is also called the **transfer function** of the system. The realization problem thus has the interpretation to construct a system with given input and output characteristics. Minimal realizations accomplish this with a state space of the smallest possible size. The system is **observable** if the only $h \in \mathfrak{H}$ such that $GT^n h = 0$ for all $n \geq 0$ is $h = 0$, and **controllable** if the span of $\operatorname{ran} T^n F$ for all $n \geq 0$ is $\mathfrak{H}$. Observe that while "observable" is equivalent to "closely outer connected," and "controllable" is equivalent to "closely inner connected" for finite-dimensional spaces, the term "closely connected" as defined in §1.2 is essentially different from "observable and controllable."

The basic facts concerning minimal realizations may be found, for example, in Bart, Gohberg, and Kaashoek [1979], Fuhrmann [1981], Kaashoek [1996], and Kailath [1980]. We briefly recall some of these for later use.

Let $S(z)$ be a rational function with values in $\mathfrak{L}(\mathfrak{F}, \mathfrak{G})$, where $\mathfrak{F}$ and $\mathfrak{G}$ are finite-dimensional Kreĭn spaces. The **local degree** of $S(z)$ at $\lambda \in \mathbf{C}$ is the rank $\delta(S; \lambda)$ of the operator matrix

$$\begin{pmatrix} S_{-q} & S_{-q+1} & S_{-q+2} & \cdots & S_{-1} \\ 0 & S_{-q} & S_{-q+1} & \cdots & S_{-2} \\ & & \cdots & & \\ 0 & 0 & 0 & \cdots & S_{-q} \end{pmatrix},$$

where

$$S(z) = \sum_{n=-q}^{\infty} S_n (z - \lambda)^n$$

in a neighborhood of λ; we define $\delta(S; \infty)$ to be the local degree of $S(1/z)$ at the origin. The **McMillan degree** of $S(z)$ is

$$\delta(S) = \sum \delta(S, \lambda),$$

where the sum is over all λ in the extended complex plane (or, what is the same thing, over the poles of $S(z)$). The definition immediately yields that

$$\delta(S) = \delta(\tilde{S}),$$

where $\tilde{S}(z) = S(\bar{z})^*$. McMillan degree is also unchanged under nonconstant linear fractional transformations of the independent variable.

Assume now that $S(z)$ is holomorphic at the origin. Then a minimal realization of $S(z)$ always exists, and the dimension of the state space is the McMillan degree of $S(z)$. Minimal realizations are both controllable and observable. Conversely, if a realization has a finite-dimensional state space and is both controllable and observable, then it is minimal. Minimal realizations are unique up to invertible mappings between state spaces. Suppose that $S(z) = \Theta_V(z)$ is a minimal realization with V given by (3.5.1). Let X be an invertible transformation from $\mathfrak{H}$ onto a space $\mathfrak{H}'$, and define V' in terms of the operators

$$T' = XTX^{-1}, \qquad F' = XF,$$
$$G' = GX^{-1}, \qquad H' = H.$$

Then $S(z) = \Theta_{V'}(z)$ is a second minimal realization. Any two minimal realizations are connnected by some such invertible X between their state spaces. Notice that the notion of equivalence of colligations defined in §2.1 is stronger: it requires that state spaces are connected by a Kreĭn space isomorphism.

If the existence of minimal realizations is granted, it is easy to exhibit a concrete instance. Suppose that a minimal realization is given by

$$S(z) = H_0 + zG_0(1 - zT_0)^{-1}F_0,$$

$$\begin{pmatrix} T_0 & F_0 \\ G_0 & H_0 \end{pmatrix} : \begin{pmatrix} \mathfrak{H}_0 \\ \mathfrak{F} \end{pmatrix} \to \begin{pmatrix} \mathfrak{H}_0 \\ \mathfrak{G} \end{pmatrix}.$$

Then for any $f \in \mathfrak{F}$,

$$\frac{S(z) - S(w)}{z - w} f = G_0(1 - zT_0)^{-1}(1 - wT_0)^{-1}F_0 f$$

for all w, z in a set Ω_S that consists of all complex numbers with a finite number of points deleted so that the formula makes sense. It is not hard to see from this that the linear mapping defined by

$$X : (1 - wT_0)^{-1}F_0 f \to \frac{S(z) - S(w)}{z - w} f, \qquad f \in \mathfrak{F}, \ w \in \Omega_S,$$

is a bijection from $\mathfrak{H}_0$ onto the span $\mathfrak{H}_S$ of functions on Ω_S of the form

$$h(z) = \frac{S(z) - S(w)}{z - w} f, \qquad f \in \mathfrak{F}, \ w \in \Omega_S.$$

The construction of X uses the fact that a minimal realization is controllable and observable. Replacing $\mathfrak{H}_0$ by $\mathfrak{H}_S$ by means of this correspondence, we then obtain a minimal realization

$$S(z) = H + zG(1 - zT)^{-1}F,$$

$$\begin{pmatrix} T & F \\ G & H \end{pmatrix} : \begin{pmatrix} \mathfrak{H}_S \\ \mathfrak{F} \end{pmatrix} \to \begin{pmatrix} \mathfrak{H}_S \\ \mathfrak{G} \end{pmatrix},$$

with state space $\mathfrak{H}_S$ and operators T, F, G, H given by (3.2.4). The inner product of $\mathfrak{H}_0$ is also transferred to $\mathfrak{H}_S$ by the correspondence, but it plays no role in the present setting. In particular, the McMillan degree of $S(z)$ is given by $\delta(S) = \dim \mathfrak{H}_S$.

In general, the spaces $\mathfrak{H}_S$ and $\mathfrak{H}(S)$ bear little resemblance to each other.

EXAMPLE. *Let $S(z) = cz$ with $\mathfrak{F} = \mathfrak{G} = \mathbf{C}$, the complex numbers in the Euclidean metric. Then a minimal realization $S(z) = \Theta_V(z)$ is obtained with $\mathfrak{H}_S = \mathbf{C}$ or $\mathfrak{H}_S = \{0\}$ according as $c \neq 0$ or $c = 0$, and*

$$(3.5.3) \qquad\qquad V = \begin{pmatrix} T & F \\ G & H \end{pmatrix} = \begin{pmatrix} 0 & c \\ 1 & 0 \end{pmatrix}.$$

Moreover:

(1) *If $|c| < 1$, then $S(z) \in \mathbf{S}_0(\mathbf{C})$ and $\dim \mathfrak{H}(S) = \infty$.*
(2) *If $|c| = 1$, then $S(z) \in \mathbf{S}_0(\mathbf{C})$ and $\dim \mathfrak{H}(S) = 1$.*
(3) *If $|c| > 1$, then $S(z) \notin \mathbf{S}_\kappa(\mathbf{C})$ for any κ, and $\mathfrak{H}(S)$ is not defined.*

The calculation of the minimal realization is immediate. If $|c| \leq 1$, $S(z)$ is holomorphic and bounded by one in the unit disk and hence belongs to $\mathbf{S}_0(\mathbf{C})$. The kernel

$$K_S(w, z) = \frac{1 - |c|^2 z\bar{w}}{1 - z\bar{w}} = 1 + (1 - |c|^2)\frac{z\bar{w}}{1 - z\bar{w}}$$

is nonnegative and the reproducing kernel for an infinite-dimensional Hilbert space $\mathfrak{H}(S)$ when $|c| < 1$ and a one-dimensional space of constant functions when $|c| = 1$. When $|c| > 1$, $K_S(w, z)$ does not have a finite number of negative squares, so $\mathfrak{H}(S)$ is not defined and $S(z)$ does not belong to $\mathbf{S}_\kappa(\mathbf{C})$ for any κ.

In the case of rational unitary functions, the notions of minimal realization and canonical coisometric, isometric, and unitary realizations essentially coincide.

THEOREM 3.5.3. *Let $\mathfrak{F}$ and $\mathfrak{G}$ be finite-dimensional Kreĭn spaces such that $\mathrm{ind}_- \mathfrak{F} = \mathrm{ind}_- \mathfrak{G}$.*

(1) *If $S(z)$ satisfies the equivalent conditions (1) and (2) in Theorem 3.5.1, then the canonical coisometric realization of $S(z)$ is minimal, and so $\dim \mathfrak{H}(S) = \delta(S)$ is the McMillan degree of $S(z)$.*

(2) *If also $S(z)^* S(z) = 1$ on the circle $|z| = 1$ excluding poles, then the canonical coisometric, isometric, and unitary realizations of $S(z)$ are all minimal, and hence the state spaces $\mathfrak{H}(S), \mathfrak{H}(\tilde{S}), \mathfrak{D}(S)$ all have the same dimension $\delta(S)$. In this case, the operators Π_1, Π_2, Λ in Theorem 3.4.1 are all unitary.*

In particular, the number κ in Theorem 3.5.1(1) satisfies $\kappa \leq \delta(S)$, but in general we cannot say how κ is determined from $S(z)$ as a rational function.

Proof. (1) By the proof of Theorem 3.5.1, if $S(z)$ satisfies the equivalent conditions, then $\dim \mathfrak{H}(S)$ is equal to the dimension of the span of functions of the form

$$\frac{\tilde{S}(1/z) - \tilde{S}(1/w)}{z - w} g$$

with $g \in \mathfrak{G}$ and w in some fixed nonempty open set. As noted above, the span of such functions can be taken as the state space of a minimal realization of $\tilde{S}(z)$, so $\dim \mathfrak{H}(S) = \delta(\tilde{S}) = \delta(S)$.

(2) Applying (1) to $\tilde{S}(z)$, we see that $\dim \mathfrak{H}(\tilde{S}) = \delta(\tilde{S}) = \delta(S)$, and so the canonical isometric realization is minimal. We use Theorem 3.4.1 to show that $\dim \mathfrak{D}(S) = \delta(S)$. Since $\dim \mathfrak{H}(S) = \dim \mathfrak{H}(\tilde{S}) = \delta(S) < \infty$, the transformation Λ defined there is onto and hence one-to-one. By (3.4.7), Π_1^* and Π_2^* have zero kernels, and hence Π_1, Π_2, Λ are all unitary. $\qquad \square$

B. Symmetry in the state spaces

We examine the effect of a symmetry of the form

$$(3.5.4) \qquad S(z) = V^* S(-z) U$$

on a function $S(z)$ belonging to a class $\mathbf{S}_\kappa(\mathfrak{F}, \mathfrak{G})$. The operators $U \in \mathfrak{L}(\mathfrak{F})$ and $V \in \mathfrak{L}(\mathfrak{G})$ are assumed to be unitary. In (3.5.4) and similar situations below, it is understood that $\Omega(S)$ is **symmetric** in the sense that $w \in \Omega(S)$ implies $-w \in \Omega(S)$.

THEOREM 3.5.4. *Suppose that $S(z) \in \mathbf{S}_\kappa(\mathfrak{F}, \mathfrak{G})$, where $\mathfrak{F}$ and $\mathfrak{G}$ are Kreĭn spaces, and that the identity (3.5.4) holds for some unitary operators $U \in \mathfrak{L}(\mathfrak{F})$ and $V \in \mathfrak{L}(\mathfrak{G})$. Then the operators*

$$(3.5.5) \qquad h(z) \to V h(-z) \qquad\qquad \text{on } \mathfrak{H}(S),$$

$$(3.5.6) \qquad k(z) \to U k(-z) \qquad\qquad \text{on } \mathfrak{H}(\tilde{S}),$$

$$(3.5.7) \qquad \begin{pmatrix} h(z) \\ k(z) \end{pmatrix} \to \begin{pmatrix} V & 0 \\ 0 & -U \end{pmatrix} \begin{pmatrix} h(-z) \\ k(-z) \end{pmatrix} \qquad \text{on } \mathfrak{D}(S),$$

are everywhere defined unitary operators in these spaces.

Proof. By (3.5.4), $\tilde{S}(z) = U^* \tilde{S}(-z) V$. Hence by the unitarity of U and V,

$$V^* K_S(-w, -z) V = \frac{1 - V^* S(-z) U U^* S(-w)^* V}{1 - z\bar{w}} = K_S(w, z),$$

$$U^* K_{\tilde{S}}(-w, -z) U = \frac{1 - U^* \tilde{S}(-z) V V^* \tilde{S}(-w)^* U}{1 - z\bar{w}} = K_{\tilde{S}}(w, z),$$

and

$$\begin{pmatrix} V^* & 0 \\ 0 & -U^* \end{pmatrix} D_S(-w, -z) \begin{pmatrix} V & 0 \\ 0 & -U \end{pmatrix}$$

$$= \begin{pmatrix} V^* K_S(-w, -z) V & -V^* \dfrac{S(-z) - S(-\bar{w})}{-z + \bar{w}} U \\[2ex] -U^* \dfrac{\tilde{S}(-z) - \tilde{S}(-\bar{w})}{-z + \bar{w}} V & U^* K_{\tilde{S}}(-w, -z) U \end{pmatrix} = D_S(w, z).$$

Let $\mathbf{R}$ be the linear relation with domain and range in $\mathfrak{H}(S)$ spanned by pairs of the form $\big(K_S(\alpha,\cdot)g, K_S(-\alpha,\cdot)Vg\big)$ with $\alpha \in \Omega(S)$ and $g \in \mathfrak{G}$. The domain and range of $\mathbf{R}$ are dense in $\mathfrak{H}(S)$. For any $\alpha_1, \alpha_2 \in \Omega(S)$ and $g_1, g_2 \in \mathfrak{G}$,

$$\langle K_S(-\alpha_1,\cdot)Vg_1, K_S(-\alpha_2,\cdot)Vg_2 \rangle_{\mathfrak{H}(S)} = \langle V^* K_S(-\alpha_1, -\alpha_2)Vg_1, g_2 \rangle_{\mathfrak{G}}$$

$$= \langle K_S(\alpha_1,\cdot)g_1, K_S(\alpha_2,\cdot)g_2 \rangle_{\mathfrak{H}(S)}$$

by the first of the three identities above. By Theorem 1.4.2(3), the closure of $\mathbf{R}$ is the graph of a unitary operator W on $\mathfrak{H}(S)$. If $h(z) = K_S(\alpha, z)g$ for some $\alpha \in \Omega(S)$ and $g \in \mathfrak{G}$, then

$$(Wh)(z) = K_S(-\alpha, z)Vg = VV^* K_S(-\alpha, z)Vg = VK_S(\alpha, -z)g = Vh(-z).$$

It follows that W is the operator (3.5.5). The operators (3.5.6) and (3.5.7) are obtained in a similar way. $\qquad\square$

A converse result holds when $\mathfrak{F}$ and $\mathfrak{G}$ are Pontryagin spaces having the same negative index.

THEOREM 3.5.5. *Suppose that* $S(z) \in \mathbf{S}_\kappa(\mathfrak{F}, \mathfrak{G})$, *where* $\mathfrak{F}$ *and* $\mathfrak{G}$ *are Pontryagin spaces with* $\mathrm{ind}_- \mathfrak{F} = \mathrm{ind}_- \mathfrak{G}$. *Assume that the only* $f \in \mathfrak{F}$ *such that* $S(z)f \equiv 0$ *is* $f = 0$. *If* $V \in \mathcal{L}(\mathfrak{G})$ *is unitary and* (3.5.5) *is an everywhere defined and unitary operator on* $\mathfrak{H}(S)$, *then there is a unitary operator* $U \in \mathcal{L}(\mathfrak{F})$ *such that* (3.5.4) *holds.*

Proof. Let W be the operator (3.5.5). If $h(z) \in \mathfrak{H}(S)$, then for any $w \in \Omega(S)$ and $g \in \mathfrak{G}$,

$$\langle h(z), W^{-1} K_S(-w, z)Vg \rangle_{\mathfrak{H}(S)} = \langle Vh(-z), K_S(-w, z)Vg \rangle_{\mathfrak{H}(S)}$$

$$= \langle Vh(w), Vg \rangle_{\mathfrak{G}} = \langle h(w), g \rangle_{\mathfrak{G}} = \langle h(z), K_S(w, z)g \rangle_{\mathfrak{H}(S)}.$$

By the arbitrariness of $h(z)$, $W^{-1} K_S(-w, z)Vg = K_S(w, z)g$ and

$$K_S(-w, z)Vg = WK_S(w, z)g = VK_S(w, -z)g.$$

Therefore $V^* K_S(-w, z)V = K_S(w, -z)$. Hence on replacing z by $-z$, we obtain $V^* K_S(-w, -z)V = K_S(w, z)$, and so

$$V^* S(-z)S(-w)^* V = S(z)S(w)^*.$$

Define a linear relation $\mathbf{R} \subseteq \mathfrak{F} \times \mathfrak{F}$ as the span of all pairs $\big(S(-w)^*Vg, S(w)^*g\big)$, $w \in \Omega(S)$ and $g \in \mathfrak{G}$. Since $S(z)f \equiv 0$ only for $f = 0$, $\mathbf{R}$ has dense domain and range. The last identity implies that the relation is isometric. By Theorem 1.4.2(3), the closure of $\mathbf{R}$ is the graph of a unitary operator $U^* \in \mathcal{L}(\mathfrak{F})$. By construction, $U^* S(-w)^* V = S(w)^*$ for all $w \in \Omega(S)$, so (3.5.4) holds. $\qquad\square$

Symmetry can also be characterized in terms of realizations.

(1) $\Rightarrow$ (2), (4), (5), (6) Assuming (1), we have equality throughout (4.1.3). In particular, $\kappa_2' = \kappa_2$. Then (2) and (5) follow from Theorem 1.5.5, and (4) holds by Theorem 1.5.7. We check (6) using the condition for unitarity in Theorem 3.2.3(2). If $f \in \mathfrak{F}$ and $S_2(z)f \in \mathfrak{H}(S_2)$, then by (4),

$$S(z)f = S_1(z)S_2(z)f \in S_1\mathfrak{H}(S_2) \subseteq \mathfrak{H}(S).$$

Thus if V is unitary, Theorem 3.2.3(2) implies $f = 0$, and hence by the same result V_2 is unitary.

(2) $\Rightarrow$ (1) Assume (2). Since $\mathfrak{H}(S_1)$ is contained contractively in $\mathfrak{H}(S)$, the kernel

(4.1.4) $$S_1(z)K_{S_2}(w, z)S_1(w)^* = K_S(w, z) - K_{S_1}(w, z)$$

has $\kappa - \kappa_1$ negative squares by Theorem 1.5.6. Since the same kernel has κ_2 negative squares by assumption, $\kappa_2 = \kappa - \kappa_1$, that is, (1) holds.

(1) $\Rightarrow$ (3) Assume (1). Then (4) and (5) hold by the first part of the proof. By (5) and the characterization of complementary spaces in Theorem 1.5.3(2), the mapping $(h_1, k_2) \to h_1 + k_2$ is a partial isometry from $\mathfrak{H}(S_1) \times S_1\mathfrak{H}(S_2)$ onto $\mathfrak{H}(S)$ whose kernel is a Hilbert space. By (4), the mapping

$$(h_1, h_2) \to (h_1, S_1h_2)$$

is a partial isometry from $\mathfrak{H}(S_1) \times \mathfrak{H}(S_2)$ onto $\mathfrak{H}(S_1) \times S_1\mathfrak{H}(S_2)$ whose kernel is a Hilbert space. In a natural way, this allows us to write the set of elements (h_1, h_2) in $\mathfrak{H}(S_1) \times \mathfrak{H}(S_2)$ such that $h_1(z) + S_1(z)h_2(z) \equiv 0$ as the direct sum of two Hilbert spaces, and this yields (3).

(3) $\Rightarrow$ (1) Assume (3). Define $\mathfrak{H}'$ to be the set of functions of the form

$$h = h_1 + S_1h_2, \qquad (h_1, h_2) \in \mathfrak{H}(S_1) \times \mathfrak{H}(S_2).$$

Our hypotheses imply that $\mathfrak{H}'$ is a Pontryagin space in a unique inner product such that

$$(h_1, h_2) \to h_1 + S_1h_2$$

is a partial isometry from $\mathfrak{H}(S_1) \times \mathfrak{H}(S_2)$ onto $\mathfrak{H}'$ whose kernel is a Hilbert space. A straightforward argument shows that $\mathfrak{H}'$ has the reproducing kernel (4.1.2), and so $\mathfrak{H}'$ is equal isometrically to $\mathfrak{H}(S)$. Since the kernel of the partial isometry is a Hilbert space, the negative index of $\mathfrak{H}(S)$ is equal to the sum of the negative indices of $\mathfrak{H}(S_1)$ and $\mathfrak{H}(S_2)$, that is, $\kappa = \kappa_1 + \kappa_2$, which is (1). $\qquad\square$

A number of applications of Theorem 4.1.1 involving Blaschke-Potapov factors and their inverses will be given in §4.2 and the Appendix (see the Appendix for definitions). Another basic example is to multiply by z.

EXAMPLE 1. *Let $S(z)$ belong to $\mathbf{S}_\kappa(\mathfrak{F}, \mathfrak{G})$, where $\mathfrak{F}$ and $\mathfrak{G}$ have the same finite negative index ν. In particular, it is assumed that $S(z)$ is holomorphic in a neighborhood of the origin. Then $zS(z) \in \mathbf{S}_{\nu+\kappa}(\mathfrak{F}, \mathfrak{G})$.*

(1) *The factorization*

$$zS(z) = z1_\mathfrak{G} \cdot S(z)$$

satisfies the equivalent conditions of Theorem 4.1.1, and it leads to the orthogonal direct decomposition $\mathfrak{H}(zS) = \mathfrak{G} \oplus z\mathfrak{H}(S)$.

(2) *The factorization*

$$zS(z) = S(z) \cdot z1_\mathfrak{F}$$

satisfies the equivalent conditions of Theorem 4.1.1. The inclusion of $\mathfrak{H}(S)$ in $\mathfrak{H}(zS)$ is thus contractive. It is isometric if and only if $S(z)$ satisfies the equivalent conditions in Theorem 3.2.3(1).

The canonical coisometric colligations for $S(z)$ and $zS(z)$ are simultaneously unitary or not.

Here we have written $\mathfrak{H}(z1_\mathfrak{G}) = \mathfrak{G}$, identifying constants and constant functions in the usual way. To verify the assertions, notice that since $K_{z1_\mathfrak{G}}(w, z)$ is the identity operator on $\mathfrak{G}$, $z1_\mathfrak{G}$ belongs to $\mathbf{S}_\nu(\mathfrak{G})$ by Theorem 2.5.2, and $\mathfrak{H}(z1_\mathfrak{G})$ is isometrically equal to $\mathfrak{G}$ with the interpretation that constants and constant functions are identified. Condition (3) of Theorem 4.1.1 is satisfied: if (h_1, h_2) belongs to $\mathfrak{H}(z1_\mathfrak{G}) \times \mathfrak{H}(S)$ and $h_1(z) + zh_2(z) \equiv 0$, then $h_1(z) \equiv g$ for some constant $g \in \mathfrak{G}$ and so (h_1, h_2) is the zero element of the product space. Hence $zS(z) \in \mathbf{S}_{\nu+\kappa}(\mathfrak{F}, \mathfrak{G})$, and the assertions of (1) are easily derived.

The factorization $zS(z) = S(z) \cdot z1_\mathfrak{F}$ satisfies condition (1) in Theorem 4.1.1 since $S(z) \in \mathbf{S}_\kappa(\mathfrak{F}, \mathfrak{G})$ by assumption and $zS(z) \in \mathbf{S}_{\nu+\kappa}(\mathfrak{F}, \mathfrak{G})$ and $z1_\mathfrak{F} \in \mathbf{S}_\nu(\mathfrak{F})$ by what is shown above. The inclusions of $\mathfrak{H}(S)$ and $S\mathfrak{H}(z1_\mathfrak{F})$ are isometric if and only if their intersection contains only the zero element (see Theorem 1.5.3), and this is the first of the two conditions in Theorem 3.2.3(1).

The last assertion in the example follows from Theorem 3.2.3(2), because if $f \in \mathfrak{F}$, then $S(z)f \in \mathfrak{H}(S)$ if and only if $zS(z)f \in \mathfrak{H}(zS)$.

There is a compatibility question associated with any factorization (4.1.1) which satisfies the equivalent conditions of Theorem 4.1.1. To state the question, we reformulate some of the information in Theorem 4.1.1 in another way. As noted in the proof of the theorem, the mapping

$$(h_1, h_2) \to h_1 + S_1 h_2$$

is a continuous partial isometry from $\mathfrak{H}(S_1) \times \mathfrak{H}(S_2)$ onto $\mathfrak{H}(S)$ whose kernel is a Hilbert space. Put differently, if $h_1 \in \mathfrak{H}(S_1)$ and $h_2 \in \mathfrak{H}(S_2)$, then

$$h = h_1 + S_1 h_2$$

belongs to $\mathfrak{H}(S)$, and

$$\langle h, h \rangle_{\mathfrak{H}(S)} \leq \langle h_1, h_1 \rangle_{\mathfrak{H}(S_1)} + \langle h_2, h_2 \rangle_{\mathfrak{H}(S_2)}.$$

Every $h \in \mathfrak{H}(S)$ has a unique such representation for which equality holds, and we call this representation the **minimal decomposition** of h relative to the spaces $\mathfrak{H}(S)$, $\mathfrak{H}(S_1)$, $\mathfrak{H}(S_2)$. Kernel functions give examples: the decomposition

$$K_S(w, z)g = K_{S_1}(w, z)g + S_1(z)K_{S_2}(w, z)S_1(w)^*g$$

is minimal for every $w \in \Omega(S)$ and $g \in \mathfrak{G}$ (see (4.1.2)). The compatibility question is if new minimal decompositions can be constructed from old ones by applying the main transformations in the spaces $\mathfrak{H}(S)$, $\mathfrak{H}(S_1)$, $\mathfrak{H}(S_2)$?

EXAMPLE 2. *Assume that the factorization (4.1.1) satisfies the equivalent conditions of Theorem 4.1.1 and that the canonical coisometric colligations associated with $\mathfrak{H}(S)$ and $\mathfrak{H}(S_1)$ are unitary. Then if $h = h_1 + S_1 h_2$ is a minimal decomposition, so is*

$$\frac{h(z) - h(0)}{z} = \left[\frac{h_1(z) - h_1(0)}{z} + \frac{S_1(z) - S_1(0)}{z} h_2(0) \right] + S_1(z) \frac{h_2(z) - h_2(0)}{z}.$$

To see this, first notice that the unitarity of the canonical coisometric colligation associated with $\mathfrak{H}(S)$ implies the same for $\mathfrak{H}(S_2)$ by Theorem 4.1.1(6). Thus the formulas in Theorem 3.2.5 hold in each of the spaces $\mathfrak{H}(S)$, $\mathfrak{H}(S_1)$, $\mathfrak{H}(S_2)$. These may be used in a routine way to reduce the identity

$$\left\langle \frac{h(z) - h(0)}{z}, \frac{h(z) - h(0)}{z} \right\rangle_{\mathfrak{H}(S)} = \left\langle \frac{h_1(z) - h_1(0)}{z} + \frac{S_1(z) - S_1(0)}{z} h_2(0), \right.$$

$$\left. \frac{h_1(z) - h_1(0)}{z} + \frac{S_1(z) - S_1(0)}{z} h_2(0) \right\rangle_{\mathfrak{H}(S_1)}$$

$$+ \left\langle \frac{h_2(z) - h_2(0)}{z}, \frac{h_2(z) - h_2(0)}{z} \right\rangle_{\mathfrak{H}(S_2)}$$

to

$$\langle h(z), h(z) \rangle_{\mathfrak{H}(S)} = \langle h_1(z), h_1(z) \rangle_{\mathfrak{H}(S_1)} + \langle h_2(z), h_2(z) \rangle_{\mathfrak{H}(S_2)},$$

which holds by the assumption that the decomposition $h = h_1 + S_1 h_2$ is minimal.

Of course, the conclusion in the example also holds if the only $h_1 \in \mathfrak{H}(S_1)$ and $h_2 \in \mathfrak{H}(S_2)$ such that $h_1 + S_1 h_2 \equiv 0$ are $h_1 = 0$ and $h_2 = 0$. For in this case, decompositions are unique and hence automatically minimal.

In general, it is hard to factor operator-valued functions. Sometimes factorizations can be obtained from the inclusion of spaces $\mathfrak{H}(S)$. These results are partial converses to Theorem 4.1.1.

THEOREM 4.1.2. *Suppose that $\mathfrak{H}(S_1)$ is contained contractively in $\mathfrak{H}(S)$, where $S(z)$ is in $\mathbf{S}_\kappa(\mathfrak{F}, \mathfrak{G})$ and $S_1(z)$ is in $\mathbf{S}_{\kappa_1}(\mathfrak{K}, \mathfrak{G})$ for some Pontryagin spaces $\mathfrak{F}, \mathfrak{G}, \mathfrak{K}$ having the same negative index. If $S_1(0)$ is invertible, there is an $S_2(z)$ such that (4.1.1) and the equivalent conditions in Theorem 4.1.1 hold.*

Proof. If $S_1(0)$ is invertible, then

$$S_2(z) = S_1(z)^{-1}S(z)$$

defines a holomorphic function on a neighborhood of the origin. Since we assume that $\mathfrak{H}(S_1)$ is contained contractively in $\mathfrak{H}(S)$, by Theorem 1.5.6 the kernel (4.1.4) has $\kappa - \kappa_1$ negative squares (recall that by Theorem 1.1.4, the number of negative squares of a holomorphic Hermitian kernel is independent of the region on which it is considered). Set $\kappa_2 = \kappa - \kappa_1$. Since $S_1(z)$ has invertible values in a neighborhood of the origin, $K_{S_2}(w, z)$ also has κ_2 negative squares, and thus $S_2(z)$ belongs to $\mathbf{S}_{\kappa_2}(\mathfrak{F}, \mathfrak{K})$ by Theorem 2.5.2. By construction $\kappa = \kappa_1 + \kappa_2$ and (4.1.1) and the equivalent conditions of Theorem 4.1.1 hold.   □

A deeper result is needed for what follows. Colligations are used here in an essential way.

THEOREM 4.1.3. *Assume that $S(z)$ is in $\mathbf{S}_\kappa(\mathfrak{F}, \mathfrak{G})$, where $\mathfrak{F}$ and $\mathfrak{G}$ are Pontryagin spaces such that $\mathrm{ind}_- \mathfrak{F} = \mathrm{ind}_- \mathfrak{G}$. Let $\mathfrak{P}$ be a Pontryagin subspace of $\mathfrak{H}(S)$ such that $[h(z) - h(0)]/z$ belongs to $\mathfrak{P}$ whenever $h(z)$ is in $\mathfrak{P}$. Then $\mathfrak{P}$ is isometrically equal to $\mathfrak{H}(S_1)$ for some factorization (4.1.1) which satisfies the equivalent conditions of Theorem 4.1.1.*

Proof. Write $S(z) = \Theta_V(z)$, where

$$V = \begin{pmatrix} T & F \\ G & H \end{pmatrix} : \begin{pmatrix} \mathfrak{H}(S) \\ \mathfrak{F} \end{pmatrix} \to \begin{pmatrix} \mathfrak{H}(S) \\ \mathfrak{G} \end{pmatrix}$$

is the canonical coisometric colligation in Theorem 2.2.1. Then V is closely outer connected and so

$$\tilde{S}(z) = \Theta_{V^*}(z),$$

where

$$V^* = \begin{pmatrix} T^* & G^* \\ F^* & H^* \end{pmatrix} : \begin{pmatrix} \mathfrak{H}(S) \\ \mathfrak{G} \end{pmatrix} \to \begin{pmatrix} \mathfrak{H}(S) \\ \mathfrak{F} \end{pmatrix}$$

is an isometric closely inner connected colligation. The state space of V^* has the decomposition

$$\mathfrak{H}(S) = \mathfrak{P}^\perp \oplus \mathfrak{P},$$

where $T^*\mathfrak{P}^\perp \subseteq \mathfrak{P}^\perp$. By Theorem 1.2.2, there is a factorization of the colligation in the form

$$(4.1.5) \qquad \left(\mathfrak{H}(S), \mathfrak{G}, \mathfrak{F}, V^*\right) = \left(\mathfrak{P}^\perp, \mathfrak{K}, \mathfrak{F}, V_2^*\right) \times \left(\mathfrak{P}, \mathfrak{G}, \mathfrak{K}, V_1^*\right),$$

where $\left(\mathfrak{P}^\perp, \mathfrak{K}, \mathfrak{F}, V_2^*\right)$ is unitary and $\left(\mathfrak{P}, \mathfrak{G}, \mathfrak{K}, V_1^*\right)$ is isometric. Both factors are closely inner connected by Theorem 1.2.1(2). By Theorem 1.2.3,

$$\Theta_{V^*}(z) = \Theta_{V_2^*}(z)\Theta_{V_1^*}(z),$$

and hence

$$\Theta_V(z) = \Theta_{V_1}(z)\Theta_{V_2}(z).$$

Here V_1, V_2 are coisometric and closely outer connected. In fact, V_2 is unitary and so

$$\text{ind}_-\left(\mathfrak{P}^\perp \oplus \mathfrak{K}\right) = \text{ind}_-\left(\mathfrak{P}^\perp \oplus \mathfrak{F}\right).$$

Since $\mathfrak{P}^\perp$ is a Pontryagin space, $\mathfrak{K}$ and $\mathfrak{F}$ have the same negative index, which by assumption is equal to the negative index of $\mathfrak{G}$. Thus $S(z) = S_1(z)S_2(z)$, where

$$S_1(z) = \Theta_{V_1}(z) \in \mathbf{S}_{\kappa_1}(\mathfrak{K}, \mathfrak{G}), \qquad\qquad \kappa_1 = \text{ind}_- \mathfrak{P},$$

$$S_2(z) = \Theta_{V_2}(z) \in \mathbf{S}_{\kappa_2}(\mathfrak{F}, \mathfrak{K}), \qquad\qquad \kappa_2 = \text{ind}_- \mathfrak{P}^\perp,$$

by Theorem 2.1.2(1) and Theorem 2.5.2. Since

$$\kappa_1 + \kappa_2 = \text{ind}_- \mathfrak{P} + \text{ind}_- \mathfrak{P}^\perp = \text{ind}_- \mathfrak{H}(S) = \kappa,$$

we have constructed a factorization (4.1.1) such that the equivalent conditions in Theorem 4.1.1 hold.

It remains to show that $\mathfrak{H}(S_1)$ is equal isometrically to $\mathfrak{P}$. By construction, $S_1(z) = \Theta_{V_1}(z)$ where

$$V_1 = \begin{pmatrix} T_1 & F_1 \\ G_1 & H_1 \end{pmatrix} : \begin{pmatrix} \mathfrak{P} \\ \mathfrak{K} \end{pmatrix} \to \begin{pmatrix} \mathfrak{P} \\ \mathfrak{G} \end{pmatrix}$$

is coisometric and closely outer connected. Therefore by Theorem 2.1.2(1) the reproducing kernel for $\mathfrak{H}(S_1)$ is given by

$$K_{S_1}(w, z) = G_1(1 - zT_1)^{-1}(1 - \bar{w}T_1^*)^{-1}G_1^*$$

for w, z in a neighborhood of the origin. To compute the reproducing kernel for $\mathfrak{P}$, rewrite (4.1.5) as

$$\big(\mathfrak{H}(S), \mathfrak{F}, \mathfrak{G}, V\big) = \big(\mathfrak{P}, \mathfrak{K}, \mathfrak{G}, V_1\big) \times \big(\mathfrak{P}^{\perp}, \mathfrak{F}, \mathfrak{K}, V_2\big).$$

Interpreting this relation as a matrix product of the form (1.2.6), we find that

$$T_1 = T|_{\mathfrak{P}},$$
$$G_1 = G|_{\mathfrak{P}}.$$

Since T is the difference-quotient transformation, an argument in the proof of Theorem 2.2.1 shows that

$$((1 - wT_1)^{-1}h)(z) = \frac{zh(z) - wh(w)}{z - w}$$

for any h in $\mathfrak{P}$ and w, z in a neighborhood of the origin. Then since G_1 is evaluation at the origin,

$$G_1(1 - wT_1)^{-1}h = h(w),$$

and so $E_1(w) = G_1(1 - wT_1)^{-1}$ is evaluation at w on $\mathfrak{P}$. Therefore by Theorem 1.1.2, the reproducing kernel for $\mathfrak{P}$ is given by

$$K_{\mathfrak{P}}(w, z) = E_1(z)E_1(w)^* = G_1(1 - zT_1)^{-1}(1 - \bar{w}T_1^*)^{-1}G_1^*$$

for w, z in a neighborhood of the origin. Since the reproducing kernels for $\mathfrak{H}(S_1)$ and $\mathfrak{P}$ coincide, the spaces are equal isometrically by Theorem 1.1.3. □

The intermediate space $\mathfrak{K}$ in a factorization (4.1.1) can be artificially increased by adding a Hilbert space summand to $\mathfrak{K}$ (the summand needs to be a Hilbert space so that the three coefficient spaces have the same negative index). It is natural to ask if a Hilbert space summand can be cut out so as to make the intermediate space as small as possible. This can be done, and the additional properties of the factorization are important in applications.

THEOREM 4.1.3′. *In Theorem 4.1.3, the factorization (4.1.1) can be chosen so that*

(1) *the only vector $k \in \mathfrak{K}$ such that $S_1(z)k \equiv 0$ is $k = 0$, and*
(2) *if the canonical coisometric colligation associated with $S(z)$ is unitary then so are the canonical coisometric colligations associated with both $S_1(z)$ and $S_2(z)$.*

Proof. (1) Suppose that the factorization provided by Theorem 4.1.3 is $S(z) = S_1(z)S_2(z)$, where

$$S_1(z) \in \mathbf{S}_{\kappa_1}(\mathfrak{K}, \mathfrak{G}), \quad S_2(z) \in \mathbf{S}_{\kappa_2}(\mathfrak{F}, \mathfrak{K})$$

and $\mathfrak{F}, \mathfrak{G}, \mathfrak{K}$ are Pontryagin spaces having the same negative index.

By Theorem 2.5.3,

$$\mathfrak{M} = \{k \in \mathfrak{K} : S_1(z)k \equiv 0\}$$

is a Hilbert subspace of $\mathfrak{K}$. Write $\mathfrak{K} = \mathfrak{K}' \oplus \mathfrak{M}$, and let E be the natural embedding mapping from $\mathfrak{K}'$ into $\mathfrak{K}$. Then EE^* is the projection onto $\mathfrak{K}'$, and

$$S_1(z) = S_1(z)EE^*$$

by Theorem 2.5.3. Therefore

$$S(z) = S_1(z)S_2(z) = S_1(z)EE^*S_2(z) = S_1'(z)S_2'(z),$$

where $S_1'(z) = S_1(z)E$ and $S_2'(z) = E^*S_2(z)$. The identity $S_1(z) = S_1(z)EE^*$ also implies

$$K_{S_1'}(w, z) = K_{S_1}(w, z),$$

and so $S_1'(z)$ belongs to $\mathbf{S}_{\kappa_1}(\mathfrak{K}', \mathfrak{G})$ and $\mathfrak{H}(S_1')$ is isometrically equal to $\mathfrak{H}(S_1)$. In a similar way

$$K_{S_2'}(w, z) = E^* K_{S_2}(w, z)E,$$

and therefore $S_2'(z) \in \mathbf{S}_{\kappa_2'}(\mathfrak{F}, \mathfrak{K}')$ where $\kappa_2' \leq \kappa_2$. Since the original factorization is chosen to satisfy the equivalent conditions in Theorem 4.1.1, $\kappa = \kappa_1 + \kappa_2$. Hence applying Theorem 4.1.1 to the factorization $S(z) = S_1'(z)S_2'(z)$, we obtain

$$\kappa \leq \kappa_1 + \kappa_2' \leq \kappa_1 + \kappa_2 = \kappa.$$

Equality thus holds throughout. Hence the new factorization $S(z) = S_1'(z)S_2'(z)$ has all of the properties of the original factorization, and in addition $S_1'(z)k' \equiv 0$ for some $k' \in \mathfrak{K}'$ only for $k' = 0$.

(2) Suppose now that the factorization (4.1.1) is chosen as in (1) and that the canonical coisometric colligation V associated with $S(z)$ is unitary. Then the condition in Theorem 3.2.5(1) holds for the space $\mathfrak{H}(S)$. Since $\mathfrak{H}(S_1)$ is contained isometrically in $\mathfrak{H}(S)$ and $S_1(z)k \equiv 0$ implies $k = 0$, the same condition holds for $\mathfrak{H}(S_1)$. Hence by Theorem 3.2.5, the canonical coisometric colligation associated with $S_1(z)$ is unitary. The canonical coisometric colligation associated with $S_2(z)$ is unitary by Theorem 4.1.1(6).   □

In many applications, the subspace $\mathfrak{P}$ in Theorems 4.1.3 and 4.1.3' is already of the form $\mathfrak{H}(S_1)$, and we want to know that this choice of $S_1(z)$ can be used in the factorization.

COROLLARY 4.1.4. *Suppose that $\mathfrak{H}(S_1)$ is contained isometrically in $\mathfrak{H}(S)$, where $S(z) \in \mathbf{S}_\kappa(\mathfrak{F}, \mathfrak{G})$, $S_1(z) \in \mathbf{S}_{\kappa_1}(\mathfrak{K}, \mathfrak{G})$, and $\mathfrak{F}, \mathfrak{G}, \mathfrak{K}$ are Pontryagin spaces having the same negative index. If the only $k \in \mathfrak{K}$ such that $S_1(z)k \equiv 0$ is $k = 0$, then*

$$S(z) = S_1(z)S_2(z),$$

with $S_2(z) \in \mathbf{S}_{\kappa_2}(\mathfrak{F}, \mathfrak{K})$, where $\kappa_2 = \kappa - \kappa_1$.

In particular, the equivalent conditions of Theorem 4.1.1 hold. The isometric inclusion of $\mathfrak{H}(S_1)$ in $\mathfrak{H}(S)$ insures that $\kappa_1 \leq \kappa$.

Proof. Notice that $\mathfrak{H}(S_1)$ is closed in $\mathfrak{H}(S)$ since both are reproducing kernel Pontryagin spaces. We apply Theorem 4.1.3' to the subspace $\mathfrak{P} = \mathfrak{H}(S_1)$ of $\mathfrak{H}(S)$. It follows that

$$S(z) = S_1'(z)S_2'(z),$$

where $S_1'(z) \in \mathbf{S}_{\kappa_1}(\mathfrak{K}', \mathfrak{G})$, $S_2'(z) \in \mathbf{S}_{\kappa_2}(\mathfrak{F}, \mathfrak{K}')$, $\kappa_2 = \kappa - \kappa_1$, $\mathfrak{K}'$ is a Pontryagin space having the same negative index as $\mathfrak{F}, \mathfrak{G}, \mathfrak{K}$, the only $k' \in \mathfrak{K}'$ such that $S_1'(z)k' \equiv 0$ is $k' = 0$, and $\mathfrak{P} = \mathfrak{H}(S_1')$. Since then $\mathfrak{H}(S_1)$ and $\mathfrak{H}(S_1')$ are equal isometrically, by Theorem 3.1.3,

$$S_1'(z) = S_1(z)W$$

for a constant unitary operator $W \in \mathfrak{L}(\mathfrak{K}', \mathfrak{K})$. The required factorization is obtained with $S_2(z) = WS_2'(z)$. □

B. Inclusion of spaces $\mathfrak{D}(S)$

It is natural to look for analogous results for the spaces $\mathfrak{D}(S)$. We shall not pursue this systematically, but we give one theorem to show the form that decompositions take in the case that $\kappa = \kappa_1 + \kappa_2$.

Let $S(z)$ belong to $\mathbf{S}_\kappa(\mathfrak{F}, \mathfrak{G})$ and have the form (4.1.1), where $\mathfrak{F}, \mathfrak{G}, \mathfrak{K}$ are Pontryagin spaces having the same negative index. Set

$$(4.1.6) \qquad R_1(z) = \begin{pmatrix} S_1(z) & 0 \\ 0 & 1_{\mathfrak{F}} \end{pmatrix}, \qquad R_2(z) = \begin{pmatrix} 1_{\mathfrak{G}} & 0 \\ 0 & S_2(z) \end{pmatrix}.$$

By direct calculation,

$$(4.1.7) \qquad D_S(w, z) = \tilde{R}_2(z)D_{S_1}(w, z)\tilde{R}_2(w)^* + R_1(z)D_{S_2}(w, z)R_1(w)^*.$$

This representation of $D_S(w, z)$ is a counterpart to (4.1.2), and in a similar way it allows us to apply the general results of §1.5 B.

THEOREM 4.1.5. *In the preceding situation, assume that* $\kappa = \kappa_1 + \kappa_2$.

(1) *The Pontryagin spaces* $\tilde{R}_2\mathfrak{D}(S_1)$ *and* $R_1\mathfrak{D}(S_2)$ *whose reproducing kernels are the summands on the right of (4.1.7) are contained contractively in* $\mathfrak{D}(S)$ *as complementary spaces in the sense of de Branges.*

(2) *Multiplication by* $\tilde{R}_2(z), R_1(z)$ *act as continuous partial isometries* W_2, W_1 *from the spaces* $\mathfrak{D}(S_1), \mathfrak{D}(S_2)$ *onto* $\tilde{R}_2\mathfrak{D}(S_1), R_1\mathfrak{D}(S_2)$, *respectively. The kernels of these partial isometries are Hilbert spaces.*

(3) *Let* A, A_1, A_2 *be the main transformations in the canonical unitary colligations for* $S(z), S_1(z), S_2(z)$. *Then the space* $\tilde{R}_2\mathfrak{D}(S_1)$ *is mapped by* A *into itself and*

$$(4.1.8) \qquad A \begin{pmatrix} h \\ k \end{pmatrix} = W_2 A_1 W_2^* \begin{pmatrix} h \\ k \end{pmatrix}, \qquad \begin{pmatrix} h \\ k \end{pmatrix} \in \tilde{R}_2\mathfrak{D}(S_1),$$

and the space $R_1\mathfrak{D}(S_2)$ *is mapped by* A^* *into itself and*

$$(4.1.9) \qquad A^* \begin{pmatrix} h \\ k \end{pmatrix} = W_1 A_2^* W_1^* \begin{pmatrix} h \\ k \end{pmatrix}, \qquad \begin{pmatrix} h \\ k \end{pmatrix} \in R_1\mathfrak{D}(S_2).$$

In particular, when the spaces $\tilde{R}_2\mathfrak{D}(S_1)$ *and* $R_1\mathfrak{D}(S_2)$ *are contained isometrically in* $\mathfrak{D}(S)$, *that is, when they intersect only in the zero element, they are closed invariant subspaces for* A *and* A^*, *respectively.*

Proof. By (4.1.7) and Theorem 1.5.5,

$$\kappa = \text{sq}_- D_S(w, z)$$
$$\leq \text{sq}_- \tilde{R}_2(z) D_{S_1}(w, z) \tilde{R}_2(w)^* + \text{sq}_- R_1(z) D_{S_2}(w, z) R_1(w)^* \leq \kappa_1 + \kappa_2.$$

Since $\kappa = \kappa_1 + \kappa_2$ by hypothesis, equality holds throughout, and (1) follows from Theorem 1.5.5. We also see that

$$\text{ind}_- \tilde{R}_2\mathfrak{D}(S_1) = \kappa_1 = \text{ind}_- \mathfrak{D}(S_1),$$
$$\text{ind}_- R_1\mathfrak{D}(S_2) = \kappa_2 = \text{ind}_- \mathfrak{D}(S_2).$$

Thus (2) follows from Theorem 1.5.7.

To check (4.1.8), write

$$\begin{pmatrix} h \\ k \end{pmatrix} = W_2 \begin{pmatrix} u \\ v \end{pmatrix}$$

with $\begin{pmatrix} u \\ v \end{pmatrix} \in \mathfrak{D}(S_1)$ in the initial space of W_2. The identity reduces to

$$AW_2 \begin{pmatrix} u \\ v \end{pmatrix} = W_2 A_1 \begin{pmatrix} u \\ v \end{pmatrix},$$

which is immediate from the definition of A in (2.3.1) and its counterpart for A_1. A similar argument yields (4.1.9), and (3) follows. □

A theory of minimal decompositions follows from Theorem 4.1.5. Assume that (4.1.1) is a factorization for which $\kappa = \kappa_1 + \kappa_2$. Theorem 4.1.5 then gives a decomposition of $\mathfrak{D}(S)$. In fact, it is not hard to see that the mapping

$$\left(\begin{pmatrix} h_1 \\ k_1 \end{pmatrix}, \begin{pmatrix} h_2 \\ k_2 \end{pmatrix} \right) \rightarrow \begin{pmatrix} h \\ k \end{pmatrix}$$

defined by setting

(4.1.10) $$\begin{pmatrix} h \\ k \end{pmatrix} = \begin{pmatrix} 1_{\mathfrak{G}} & 0 \\ 0 & \tilde{S}_2 \end{pmatrix} \begin{pmatrix} h_1 \\ k_1 \end{pmatrix} + \begin{pmatrix} S_1 & 0 \\ 0 & 1_{\mathfrak{F}} \end{pmatrix} \begin{pmatrix} h_2 \\ k_2 \end{pmatrix}$$

is a continuous partial isometry from $\mathfrak{D}(S_1) \times \mathfrak{D}(S_2)$ onto $\mathfrak{D}(S)$ whose kernel is a Hilbert space. For any representation (4.1.10),

(4.1.11) $$\left\langle \begin{pmatrix} h \\ k \end{pmatrix}, \begin{pmatrix} h \\ k \end{pmatrix} \right\rangle_{\mathfrak{D}(S)} \leq \left\langle \begin{pmatrix} h_1 \\ k_1 \end{pmatrix}, \begin{pmatrix} h_1 \\ k_1 \end{pmatrix} \right\rangle_{\mathfrak{D}(S_1)} + \left\langle \begin{pmatrix} h_2 \\ k_2 \end{pmatrix}, \begin{pmatrix} h_2 \\ k_2 \end{pmatrix} \right\rangle_{\mathfrak{D}(S_2)}$$

Every element of $\mathfrak{D}(S)$ has a unique representation (4.1.10) for which equality holds in (4.1.11), and we call this the **minimal decomposition** of the element relative to the spaces $\mathfrak{D}(S)$, $\mathfrak{D}(S_1)$, $\mathfrak{D}(S_2)$.

Again kernel functions give examples (see (4.1.7)), and the question arises if new minimal decompositions result from old ones on applying main transformations for the spaces $\mathfrak{D}(S)$, $\mathfrak{D}(S_1)$, $\mathfrak{D}(S_2)$? In contrast with Example 2 in §4.1 A, we obtain an affirmative answer with no additional hypotheses.

EXAMPLE 3. *Assume that the factorization (4.1.1) satisfies the equivalent conditions of Theorem 4.1.1. Then if (4.1.10) is a minimal decomposition, so are*

$$\begin{pmatrix} [h(z) - h(0)]/z \\ zk(z) - \tilde{S}(z)h(0) \end{pmatrix}$$

$$= \begin{pmatrix} 1_{\mathfrak{G}} & 0 \\ 0 & \tilde{S}_2(z) \end{pmatrix} \left[\begin{pmatrix} [h_1(z) - h_1(0)]/z \\ zk_1(z) - \tilde{S}_1(z)h_1(0) \end{pmatrix} + D_{S_1}(0, z) \begin{pmatrix} 0 \\ h_2(0) \end{pmatrix} \right]$$

$$+ \begin{pmatrix} S_1(z) & 0 \\ 0 & 1_{\mathfrak{F}} \end{pmatrix} \begin{pmatrix} [h_2(z) - h_2(0)]/z \\ zk_2(z) - \tilde{S}_2(z)h_2(0) \end{pmatrix}$$

and

$$\begin{pmatrix} zh(z) - S(z)k(0) \\ [k(z) - k(0)]/z \end{pmatrix}$$

$$= \begin{pmatrix} 1_{\mathfrak{G}} & 0 \\ 0 & \tilde{S}_2(z) \end{pmatrix} \begin{pmatrix} zh_1(z) - S_1(z)k_1(0) \\ [k_1(z) - k_1(0)]/z \end{pmatrix}$$

$$+ \begin{pmatrix} S_1(z) & 0 \\ 0 & 1_{\mathfrak{F}} \end{pmatrix} \left[\begin{pmatrix} zh_2(z) - S_2(z)k_2(0) \\ [k_2(z) - k_2(0)]/z \end{pmatrix} + D_{S_2}(0, z) \begin{pmatrix} k_1(0) \\ 0 \end{pmatrix} \right].$$

In fact, the identities (3.4.1) and (3.4.2) hold for arbitrary spaces $\mathfrak{D}(S)$. The minimal decompositions may thus be verified by expanding self-products and using (3.4.1) and (3.4.2) to make simplifications. We omit the straightforward details.

4.2 Kreĭn-Langer factorization

A. Existence and properties

When $\mathfrak{F}$ and $\mathfrak{G}$ are Hilbert spaces, the functions in $\mathbf{S}_\kappa(\mathfrak{F}, \mathfrak{G})$ are holomorphic on the unit disk with the exception of a finite number of poles which are accounted for by inverse Blaschke-Potapov factors. See the Appendix for the definition of a Blaschke product and other terminology used here. In this section, we present an approach which is suggested by the existence of special invariant subspaces for contraction operators on Pontryagin spaces as described in Theorem 1.3.6, and the implications of the existence of such subspaces in Theorem 1.2.2.

THEOREM 4.2.1. *Let $\mathfrak{F}$ and $\mathfrak{G}$ be Hilbert spaces, and let $S(z) \in \mathbf{S}_\kappa(\mathfrak{F}, \mathfrak{G})$. Then*

$$(4.2.1) \qquad\qquad S(z) = B_\ell(z)^{-1} S_\ell(z),$$

where $B_\ell(z)$ is a Blaschke product of degree κ with values in $\mathfrak{L}(\mathfrak{G})$ and $S_\ell(z)$ is in $\mathbf{S}_0(\mathfrak{F}, \mathfrak{G})$. For any factorization with these properties, the only $g \in \mathfrak{G}$ such that $S_\ell(w)^* g = 0$ and $B_\ell(w)^* g = 0$ for some w in the unit disk is $g = 0$. Moreover

$$(4.2.2) \qquad\qquad S(z) = S_r(z) B_r(z)^{-1},$$

where $B_r(z)$ is a Blaschke product of degree κ with values in $\mathfrak{L}(\mathfrak{F})$ and $S_r(z)$ is in $\mathbf{S}_0(\mathfrak{F}, \mathfrak{G})$. For any factorization of this form, the only $f \in \mathfrak{F}$ such that $S_r(w)f = 0$ and $B_r(w)f = 0$ for some w in the unit disk is $f = 0$.

A representation (4.2.1) is called a **left Kreĭn-Langer factorization**, and a representation (4.2.2) is called a **right Kreĭn-Langer factorization**.

Proof. It is enough to prove the result for factorizations of the type (4.2.1), because the other case then follows on replacing $S(z)$ by $\tilde{S}(z)$. All assertions are clear when $\kappa = 0$, so we assume $\kappa > 0$.

Consider the canonical coisometric realization of $S(z)$ as in Theorem 2.2.1. The transformation $T : h(z) \rightarrow [h(z) - h(0)]/z$ is a bicontraction on $\mathfrak{H}(S)$ by Theorem 3.2.6 and (1.3.12), since we assume that $\mathfrak{F}$ and $\mathfrak{G}$ are Hilbert spaces.

Let λ be an eigenvalue of T. A corresponding eigenfunction is necessarily of the form

$$h(z) = g/(1 - \lambda z)$$

for some nonzero vector $g \in \mathfrak{G}$. By (3.1.1),

$$|\lambda|^2 \langle h, h \rangle_{\mathfrak{H}(S)} \leq \langle h, h \rangle_{\mathfrak{H}(S)} - \langle h(0), h(0) \rangle_{\mathfrak{G}},$$

and so

(4.2.3) $$0 < \|g\|_{\mathfrak{G}}^2 \leq (1 - |\lambda|^2)\langle h, h \rangle_{\mathfrak{H}(S)}.$$

In particular, T has no eigenvalue having a neutral eigenfunction and no eigenvalue of unit modulus. By Theorem 1.3.6(4), T has a unique κ-dimensional nonpositive invariant subspace $\mathfrak{N}$.

Let $\beta_1 \in \sigma(T|_{\mathfrak{N}})$. Then β_1 is an eigenvalue, $|\beta_1| > 1$, and any eigenfunction h_1 belongs to $\mathfrak{N}$ and satisfies $\langle h_1, h_1 \rangle_{\mathfrak{H}(S)} < 0$ by (4.2.3) applied with $\lambda = \beta_1$ and $h = h_1$. We may choose this eigenfunction of the form

$$h_1(z) = g_1/(1 - \beta_1 z), \qquad g_1 \in \mathfrak{G}, \ \|g_1\| = 1.$$

Put

$$B_1(z) = 1 - P_1 + \frac{z - \alpha_1}{1 - \bar{\alpha}_1 z} P_1,$$

where $\alpha_1 = 1/\beta_1$ and $P_1 = \langle \cdot, g_1 \rangle_{\mathfrak{G}} g_1$ is the projection onto the span of g_1 in $\mathfrak{G}$. We show that $\mathfrak{H}(B_1^{-1})$ is contained contractively in $\mathfrak{H}(S)$. According to Example 1 in the Appendix, $\mathfrak{H}(B_1^{-1})$ is spanned by the function

$$h_0(z) = \frac{\sqrt{1 - |\alpha_1|^2}}{z - \alpha_1} g_1 = -\frac{\beta_1}{|\beta_1|} \frac{\sqrt{|\beta_1|^2 - 1}}{1 - \beta_1 z} g_1,$$

and $\langle h_0, h_0 \rangle_{\mathfrak{H}(B_1^{-1})} = -1$. Since $h_0 = c\sqrt{|\beta_1|^2 - 1}\, h_1$ with $|c| = 1$,

$$\langle h_1, h_1 \rangle_{\mathfrak{H}(B_1^{-1})} = -1/(|\beta_1|^2 - 1).$$

Hence by (4.2.3) with $h = h_1$,

$$\langle h_1, h_1 \rangle_{\mathfrak{H}(S)} \leq \frac{-1}{|\beta_1|^2 - 1} = \langle h_1, h_1 \rangle_{\mathfrak{H}(B_1^{-1})},$$

showing that $\mathfrak{H}(B_1^{-1})$ is contained contractively in $\mathfrak{H}(S)$. By Theorem 4.1.2,

$$S(z) = B_1(z)^{-1} S_1(z),$$

where $S_1(z) \in \mathbf{S}_{\kappa-1}(\mathfrak{F}, \mathfrak{G})$. Iterating this construction, we obtain a factorization (4.2.1) where $B_\ell(z)$ is a Blaschke product of degree κ with values in $\mathfrak{L}(\mathfrak{G})$ and $S_\ell(z)$ is in $\mathbf{S}_0(\mathfrak{F}, \mathfrak{G})$.

Consider any factorization of the form (4.2.1). Suppose that $|w| < 1$, $g \in \mathfrak{G}$, $S_\ell(w)^* g = 0$, and $B_\ell(w)^* g = 0$. If $g \neq 0$, then by Example 1(1) of the Appendix, there is a simple Blaschke-Potapov factor $B_0(z)$ such that $\mathfrak{H}(B_0)$ is equal isometrically to the span of $h(z) = g/(1 - \bar{w}z)$ in the inner product of $H_{\mathfrak{G}}^2$, that is,

$$\|h\|_{\mathfrak{H}(B_0)}^2 = \|g\|_{\mathfrak{G}}^2 /(1 - |w|^2).$$

Our hypotheses imply that

$$h(z) = K_{B_\ell}(w, z)g = K_{S_\ell}(w, z)g,$$

and from these relations we find that $\mathfrak{H}(B_0)$ is contained isometrically in both $\mathfrak{H}(B_\ell)$ and $\mathfrak{H}(S_\ell)$. By Corollary 4.1.4,

$$B_\ell(z) = B_0(z)B_\ell'(z),$$
$$S_\ell(z) = B_0(z)S_\ell'(z),$$

where $B_\ell'(z) \in \mathbf{S}_0(\mathfrak{G})$, $S_\ell'(z) \in \mathbf{S}_\kappa(\mathfrak{F}, \mathfrak{G})$, and both factorizations satisfy the equivalent conditions of Theorem 4.1.1; in particular, the canonical coisometric colligation for $B_\ell'(z)$ is unitary by Theorem 4.1.1(6). Since the inclusion of $\mathfrak{H}(B_0)$ in $\mathfrak{H}(B_\ell)$ is isometric and the values of $B_0(z)$ are invertible except at one point, multiplication by $B_0(z)$ is an isometry from $\mathfrak{H}(B_\ell')$ onto the orthogonal complement of $\mathfrak{H}(B_0)$ in $\mathfrak{H}(B_\ell)$, and

$$\mathfrak{H}(B_\ell) = \mathfrak{H}(B_0) \oplus B_0 \mathfrak{H}(B_\ell').$$

Hence by Theorem A2 (Appendix), $B_\ell'(z)$ is a Blaschke product of degree $\kappa - 1$. But now

$$S(z) = B_\ell(z)^{-1} S_\ell(z) = B_\ell'(z)^{-1} S_\ell'(z),$$

and hence, by Theorem A3 (Appendix) and Theorem 4.1.1, $S(z)$ is in $\mathbf{S}_{\kappa'}(\mathfrak{F}, \mathfrak{G})$ for some $\kappa' \leq \kappa - 1$, a contradiction. Therefore $g = 0$, and the last statement of the theorem is proved. $\square$

COROLLARY 4.2.2. *If $\mathfrak{F}$ and $\mathfrak{G}$ are Hilbert spaces, every $S(z)$ in $\mathbf{S}_\kappa(\mathfrak{F}, \mathfrak{G})$ has a holomorphic extension to the unit disk $\mathbf{D}$ except for at most κ nonzero points where the function has poles.*

Proof. Theorem 4.2.1 reduces the assertion to the case $\kappa = 0$, and then the result is given by Theorem 2.5.5. $\square$

Left and right Kreĭn-Langer factorizations are well behaved as examples of factorizations of the form (4.1.1), and they have special properties.

THEOREM 4.2.3. *Assume that* $\mathfrak{F}$ *and* $\mathfrak{G}$ *are Hilbert spaces and that* $S(z)$ *belongs to* $\mathbf{S}_\kappa(\mathfrak{F}, \mathfrak{G})$.

(1) *Arbitrary left and right Kreĭn-Langer factorizations* (4.2.1) *and* (4.2.2) *satisfy the equivalent conditions of Theorem* 4.1.1.

(2) *For any left Kreĭn-Langer factorization* (4.2.1), *multiplication by the function* $B_\ell(z)^{-1}$ *maps* $\mathfrak{H}(S_\ell)$ *isometrically onto* $B_\ell^{-1}\mathfrak{H}(S_\ell)$. *For any right Kreĭn-Langer factorization* (4.2.2), *multiplication by* $S_r(z)$ *maps* $\mathfrak{H}(B_r^{-1})$ *isometrically onto* $S_r\mathfrak{H}(B_r^{-1})$.

(3) *The contractive transformation*

$$T : h(z) \to [h(z) - h(0)]/z$$

in $\mathfrak{H}(S)$ *has a unique* κ-*dimensional nonpositive invariant subspace. It is uniformly negative in the inner product of* $\mathfrak{H}(S)$ *and coincides with* $\mathfrak{H}(B_\ell^{-1})$ *as a vector space for any left Kreĭn-Langer factorization* (4.2.1).

(4) *If* (4.2.2) *is a right Kreĭn-Langer factorization,* $\mathfrak{H}(S_r)$ *and* $S_r\mathfrak{H}(B_r^{-1})$ *are contained isometrically in* $\mathfrak{H}(S)$ *and*

$$\mathfrak{H}(S) = \mathfrak{H}(S_r) \oplus S_r\mathfrak{H}(B_r^{-1})$$

in the usual sense of orthogonal direct sums. The adjoint T^* *of the transformation in* (3) *has a unique* κ-*dimensional nonpositive invariant subspace, and this subspace is uniformly negative and coincides with* $S_r\mathfrak{H}(B_r^{-1})$.

Proof. (1) By Theorem A3 (Appendix), any left or right Kreĭn-Langer factorization satisfies condition (1) in Theorem 4.1.1: the relation $\kappa = \kappa_1 + \kappa_2$ in Theorem 4.1.1(1) has the form $\kappa = \kappa + 0$ in the case (4.2.1) and $\kappa = 0 + \kappa$ in the case (4.2.2).

(2) For a factorization (4.2.1), multiplication by $B_\ell(z)^{-1}$ is a continuous partial isometry from $\mathfrak{H}(S_\ell)$ onto $B_\ell^{-1}\mathfrak{H}(S_\ell)$ by Theorem 4.1.1(4). Multiplication by $B_\ell(z)^{-1}$ clearly has zero kernel, so the partial isometry is an isometry. Again by Theorem 4.1.1(4), for a factorization (4.2.2), multiplication by $S_r(z)$ is a continuous partial isometry from $\mathfrak{H}(B_r^{-1})$ onto $S_r\mathfrak{H}(B_r^{-1})$, and the kernel of the partial isometry is a Hilbert space. Since $\mathfrak{H}(B_r^{-1})$ is the antispace of a Hilbert space (Theorem A3, Appendix), the kernel is zero.

(3) It is already shown in the proof of Theorem 4.2.1 that T has a unique κ-dimensional nonpositive invariant subspace $\mathfrak{N}$. By Theorem A3 (Appendix), $\mathfrak{H}(B_\ell^{-1})$ is the antispace of a κ-dimensional Hilbert space. By part (1) of the

theorem, $\mathfrak{H}(B_\ell^{-1})$ is contained contractively in $\mathfrak{H}(S)$ and hence, when viewed as a subspace of $\mathfrak{H}(S)$, is an invariant subspace for T. For any $h \in \mathfrak{H}(B_\ell^{-1})$,

$$\langle h, h \rangle_{\mathfrak{H}(S)} \leq \langle h, h \rangle_{\mathfrak{H}(B_\ell^{-1})} \leq 0,$$

with strict inequality on the right if $h \neq 0$. Thus $\mathfrak{H}(B_\ell^{-1})$ is uniformly negative in the inner product of $\mathfrak{H}(S)$. In particular $\mathfrak{N} = \mathfrak{H}(B_\ell^{-1})$ as subspaces of $\mathfrak{H}(S)$, and the assertions of (3) follow.

(4) The first statement in (4) will follow from properties of contractively contained spaces if we can show that $\mathfrak{H}(S_r)$ and $S_r\mathfrak{H}(B_r^{-1})$ have only the zero element in common (see Theorem 4.1.1(5) and Theorem 1.5.3(4)). To do this, choose an algebraic basis $f_1(z), \ldots, f_\kappa(z)$ for $\mathfrak{H}(B_r^{-1})$ consisting of chains of the form (A5) as in Theorem A4 (Appendix). Then $S_r\mathfrak{H}(B_r^{-1})$ is spanned by the functions $S_r(z)f_1(z), \ldots, S_r(z)f_\kappa(z)$. Suppose that $h(z)$ belongs to both $\mathfrak{H}(S_r)$ and $S_r\mathfrak{H}(B_r^{-1})$. Then it has a representation

$$h(z) = S_r(z) \sum_{j=1}^\kappa \eta_j f_j(z)$$

for some numbers $\eta_1, \ldots, \eta_\kappa$. If $h(z) \not\equiv 0$, then the part $\sum_{j=1}^\kappa \eta_j f_j(z)$ in $\mathfrak{H}(B_r^{-1})$ has a pole, say α. Necessarily, α is a null point of $B_r(z)$, and we have a Laurent expansion

$$\sum_{j=1}^\kappa \eta_j f_j(z) = \frac{c_n}{(1 - z/\alpha)^{n+1}} + \frac{c_{n-1}}{(1 - z/\alpha)^n} + \cdots$$

in a deleted neighborhood of α with n some nonnegative integer and coefficients $c_n, c_{n-1}, \ldots$ in $\mathfrak{F}$ such that $c_n \neq 0$. The Laurent expansion may be computed from the expansions of the separate terms of the chains (A5). The leading term $c_n/(1 - z/\alpha)^{n+1}$ must come from one or several such chains, and so

$$c_n = \sum_{j=1}^p \xi_j v_{0j},$$

where $v_{01}, \ldots, v_{0p}$ are the vectors that play the role of v_0 in (A5) in those chains which contribute to the leading term. Since $h(z)$ also belongs to $\mathfrak{H}(S_r)$ and $S_r(z)$ is in $\mathbf{S}_0(\mathfrak{F}, \mathfrak{G})$, $h(z)$ is analytic at α by Corollary 4.2.2. This is only possible if

(4.2.4) $$S_r(\alpha)c_n = 0.$$

Since also $B_r(\alpha)v_{0j} = 0$ for all $j = 1, \ldots, p$ by the choice of the chains in Theorem A4 (Appendix),

(4.2.5) $$B_r(\alpha)c_n = 0.$$

The relations (4.2.4) and (4.2.5) imply that $c_n = 0$, a contradiction. Therefore $h(z) \equiv 0$ and the first statement in (4) follows.

The main step in proving the second statement in (4) is to show that T^* has a unique κ-dimensional nonpositive invariant subspace. By Theorem 1.3.6(4), it is enough to show that T^* has no eigenvalue λ with eigenfunction h such that $|\lambda| \leq 1$ and $\langle h, h \rangle_{\mathfrak{H}(S)} = 0$. Suppose to the contrary that λ is such an eigenvalue and h is such an eigenfunction. Adapting an argument in Iokhvidov, Kreĭn, and Langer [1982] (Lemma 11.5), we define a nonnegative inner product on $\mathfrak{H}(S)$ by

$$\langle h_1, h_2 \rangle_0 = \langle (1 - TT^*)h_1, h_2 \rangle_{\mathfrak{H}(S)}, \qquad h_1, h_2 \in \mathfrak{H}(S).$$

Our assumptions imply that $\langle h, h \rangle_0 = 0$. By the generalized Schwarz inequality, then $\langle h, k \rangle_0 = 0$ for all $k \in \mathfrak{H}(S)$. This implies $(1 - TT^*)h = 0$, hence

$$h = TT^*h = \lambda T h.$$

For $\lambda = 0$ we have directly $h = 0$, contradicting the assumption that h is an eigenfunction. If λ is not zero, then $1/\lambda$ is an eigenvalue for T having a neutral eigenfunction, and no such eigenvalues exist by the proof of Theorem 4.2.1. Thus T^* satisfies the condition in Theorem 1.3.6(4), and so T^* has a unique κ-dimensional nonpositive invariant subspace.

To see that this space is $S_r \mathfrak{H}(B_r^{-1})$, observe: (a) this space is κ-dimensional, (b) it is nonpositive and in fact uniformly negative because multiplication by S_r is isometric on $\mathfrak{H}(B_r^{-1})$, and (c) it is invariant under T^* because its orthogonal complement $\mathfrak{H}(S_r)$ is invariant under T. $\qquad \square$

Kreĭn-Langer factorizations are essentially unique.

THEOREM 4.2.4. *Assume that $\mathfrak{F}$ and $\mathfrak{G}$ are Hilbert spaces and that $S(z)$ belongs to $\mathbf{S}_\kappa(\mathfrak{F}, \mathfrak{G})$. If (4.2.1) and $S(z) = B'_\ell(z)^{-1} S'_\ell(z)$ are two left Kreĭn-Langer factorizations, there is a unitary operator $U \in \mathfrak{L}(\mathfrak{G})$ such that*

$$(4.2.6) \qquad\qquad B'_\ell(z) = U B_\ell(z), \qquad S'_\ell(z) = U S_\ell(z).$$

If (4.2.2) and $S(z) = S'_r(z) B'_r(z)^{-1}$ are two right Kreĭn-Langer factorizations, there is a unitary operator $V \in \mathfrak{L}(\mathfrak{F})$ such that

$$(4.2.7) \qquad\qquad B'_r(z) = B_r(z) V, \qquad S'_r(z) = S_r(z) V.$$

The spaces $\mathfrak{H}(S_r)$, $S_r \mathfrak{H}(B_r^{-1})$, $\mathfrak{H}(B_\ell^{-1})$, $B_\ell^{-1} \mathfrak{H}(S_\ell)$ are independent of the choice of Kreĭn-Langer factorizations, since by (4.2.6) and (4.2.7) their reproducing kernels are identical for any choice of factorizations.

Proof. It is enough to prove this for either one of the two cases, since the other is then obtained by replacing $S(z)$ by $\tilde{S}(z)$.

Let (4.2.1) and $S(z) = B'_\ell(z)^{-1}S'_\ell(z)$ be two left Kreĭn-Langer factorizations. By Theorem 4.2.3(1), $\mathfrak{H}(B_\ell^{-1})$ and $\mathfrak{H}(B'^{-1}_\ell)$ are contained contractively in $\mathfrak{H}(S)$. By Theorem A3 (Appendix), each of these spaces is a κ-dimensional nonpositive subspace of $\mathfrak{H}(S)$ which is invariant under the transformation

$$T : h(z) \to [h(z) - h(0)]/z$$

in $\mathfrak{H}(S)$. It was shown in the proof of Theorem 4.2.1 that T has a unique κ-dimensional nonpositive invariant subspace. Hence $\mathfrak{H}(B_\ell^{-1})$ and $\mathfrak{H}(B'^{-1}_\ell)$ are equal as sets. By Theorem A5 (Appendix), $\mathfrak{H}(B_\ell^{-1})$ and $\mathfrak{H}(B'^{-1}_\ell)$ are equal isometrically, and hence, by Theorem 3.1.3,

$$B_\ell(z)^{-1} = B'_\ell(z)^{-1}U$$

for some constant unitary operator $U \in \mathfrak{L}(\mathfrak{G})$. With a little manipulation, we get (4.2.6). $\qquad\square$

A converse question has not yet been addressed. Suppose we try to construct examples by multiplying an element of $\mathbf{S}_0(\mathfrak{F}, \mathfrak{G})$ by the inverse of a Blaschke product of degree κ. Do we obtain an element of $\mathbf{S}_\kappa(\mathfrak{F}, \mathfrak{G})$? The answer is affirmative, if we eliminate the possibility of cancellation of common factors.

THEOREM 4.2.5. *Let $\mathfrak{F}$ and $\mathfrak{G}$ be Hilbert spaces.*

(1) *Define $S(z)$ by (4.2.1), where $B_\ell(z)$ is a Blaschke product of degree κ with values in $\mathfrak{L}(\mathfrak{G})$ and $S_\ell(z)$ is in $\mathbf{S}_0(\mathfrak{F}, \mathfrak{G})$. Then $S(z)$ belongs to $\mathbf{S}_{\kappa'}(\mathfrak{F}, \mathfrak{G})$ for some $\kappa' \leq \kappa$. If the only $g \in \mathfrak{G}$ such that*

$$S_\ell(w)^*g = 0 \quad and \quad B_\ell(w)^*g = 0$$

for some w in the unit disk is $g = 0$, then $\kappa' = \kappa$ and (4.2.1) is a left Kreĭn-Langer factorization.

(2) *Define $S(z)$ by (4.2.2), where $B_r(z)$ is a Blaschke product of degree κ with values in $\mathfrak{L}(\mathfrak{F})$ and $S_r(z)$ is in $\mathbf{S}_0(\mathfrak{F}, \mathfrak{G})$. Then $S(z)$ belongs to $\mathbf{S}_{\kappa'}(\mathfrak{F}, \mathfrak{G})$ for some $\kappa' \leq \kappa$. If the only $f \in \mathfrak{F}$ such that*

$$S_r(w)f = 0 \quad and \quad B_r(w)f = 0$$

for some w in the unit disk is $f = 0$, then $\kappa' = \kappa$ and (4.2.2) is a right Kreĭn-Langer factorization.

Proof. It is sufficient to prove (2), since once this is known we obtain (1) by applying (2) to the function $\tilde{S}(z)$.

Applying Theorem 4.1.1 to the factorization (4.2.2), we see that $S(z)$ belongs to $\mathbf{S}_{\kappa'}(\mathfrak{F}, \mathfrak{G})$, where

$$\kappa' \leq 0 + \kappa = \kappa.$$

Assume that the only $f \in \mathfrak{F}$ such that $S_r(w)f = 0$ and $B_r(w)f = 0$ for some w in the unit disk is $f = 0$. We show that $\kappa' = \kappa$ by verifying the third of the equivalent conditions in Theorem 4.1.1. Suppose that (h_1, h_2) belongs to $\mathfrak{H}(S_r) \times \mathfrak{H}(B_r^{-1})$ and

$$h_1(z) + S_r(z)h_2(z) \equiv 0.$$

Choose an algebraic basis $f_1(z), \dots, f_\kappa(z)$ for $\mathfrak{H}(B_r^{-1})$ consisting of chains of the form (A5) as in the proof of Theorem 4.2.3(4). Let

$$h_2(z) = \sum_{j=1}^{\kappa} \eta_j f_j(z)$$

be the representation of $h_2(z)$ in terms of this basis. An argument similar to one given in the proof of Theorem 4.2.3(4) shows that $h_2(z) \equiv 0$. Hence also $h_1(z) \equiv 0$. Condition (3) of Theorem 4.1.1 thus holds, and therefore $\kappa' = \kappa$, completing the proof of (2).   □

An example illustrates these results and resolves a question which is left open in Theorem 4.2.3. By part (1) of that result, if (4.2.1) is any left Kreĭn-Langer factorization, $\mathfrak{H}(B_\ell^{-1})$ and $B_\ell^{-1}\mathfrak{H}(S_\ell)$ are contained contractively in $\mathfrak{H}(S)$ and are complementary in the sense of de Branges (see parts (2) and (5) of Theorem 4.1.1). Are the inclusions isometric with

$$\mathfrak{H}(S) = \mathfrak{H}(B_\ell^{-1}) \oplus B_\ell^{-1}\mathfrak{H}(S_\ell)?$$

In general, this is false.

EXAMPLE. *Define $S(z)$ in* $\mathbf{S}_1(\mathfrak{F}, \mathfrak{G})$, $\mathfrak{F} = \mathbf{C}^2$, $\mathfrak{G} = \mathbf{C}$, *by*

$$S(z) = \frac{1}{\sqrt{2}} \left(a(z) \quad \frac{1}{b(z)} \right),$$

where $a(z)$ is a complex-valued holomorphic function which is bounded by one on the unit disk and $b(z) = (z - \alpha)/(1 - \bar{\alpha}z)$ with $0 < |\alpha| < 1$. Then

$$S(z) = B_\ell(z)^{-1}S_\ell(z) = b(z)^{-1} \left(\frac{1}{\sqrt{2}} a(z)b(z) \quad \frac{1}{\sqrt{2}} \right),$$

$$S(z) = S_r(z)B_r(z)^{-1} = \left(\frac{1}{\sqrt{2}} a(z) \quad \frac{1}{\sqrt{2}} \right) \begin{pmatrix} 1 & 0 \\ 0 & b(z) \end{pmatrix}^{-1}$$

has at most κ negative squares as a function with values in $\mathfrak{L}(|\mathfrak{G}_-|)$. Define a kernel whose values are operators in $\mathfrak{L}(|\mathfrak{G}_-|)$ by

$$L(w, z) = \frac{\tau S_{22}(z)\sigma^*\sigma S_{22}(w)^*\tau^* - 1_{|\mathfrak{G}_-|}}{1 - z\bar{w}}.$$

Then $L(w, z)$ has at most κ negative squares because it is the sum of a kernel having at most κ negative squares and a nonnegative kernel. By Lemma 4.4.3,

$$\tau S_{22}(z)\sigma^*\sigma S_{22}(z)^*\tau^*$$

is invertible for all but at most κ points, and hence so is $S_{22}(z)S_{22}(z)^*$. A similar argument using $K_{\tilde{S}}(w, z)$ shows that $S_{22}(z)^* S_{22}(z)$ is invertible for all but at most κ points. Thus $S_{22}(z)$ is invertible for all but at most 2κ points.

(2) The proof here is the same as for Theorem 4.3.3(2). $\qquad\square$

THEOREM 4.4.4. *Suppose that $S(z)$ is in $\mathbf{S}_\kappa(\mathfrak{F}, \mathfrak{G})$, where $\mathfrak{F}$ and $\mathfrak{G}$ are Kreĭn spaces. If $S_{22}(0)$ is invertible, then there exists a coisometric and closely outer connected colligation*

$$(4.4.1) \qquad V = \begin{pmatrix} T & F \\ G & H \end{pmatrix} : \begin{pmatrix} \mathfrak{H}(S) \\ \mathfrak{F} \end{pmatrix} \rightarrow \begin{pmatrix} \mathfrak{H}(S) \\ \mathfrak{G} \end{pmatrix}$$

such that for all h in $\mathfrak{H}(S)$, f in $\mathfrak{F}$, and $z \in \Omega(S)$,

$$(4.4.2) \qquad \begin{cases} (Th)(z) = \dfrac{h(z) - h(0)}{z}, \\ (Ff)(z) = \dfrac{S(z) - S(0)}{z}f, \\ \quad Gh = h(0), \\ \quad Hf = S(0)f, \end{cases}$$

and such that $S(z) = \Theta_V(z)$ for all $z \in \Omega(S) \cap \Omega(\Theta_V)$.

Proof. Define V by (4.4.1) and (4.4.2). The main problem is to show that V is everywhere defined and continuous. In the case of G and H this is clear.

By Theorem 4.4.1(2), the Potapov-Ginzburg transform $\Sigma(z)$ of $S(z)$ is in $\mathbf{S}_\kappa(\mathfrak{F}', \mathfrak{G}')$. Let

$$(4.4.3) \qquad V_\Sigma = \begin{pmatrix} T_\Sigma & F_\Sigma \\ G_\Sigma & H_\Sigma \end{pmatrix} : \begin{pmatrix} \mathfrak{H}(\Sigma) \\ \mathfrak{F}' \end{pmatrix} \rightarrow \begin{pmatrix} \mathfrak{H}(\Sigma) \\ \mathfrak{G}' \end{pmatrix}$$

be the canonical coisometric colligation associated with $\Sigma(z)$. Define $\Phi(z)$ and $\Psi(z)$ by (4.3.8) and (4.3.9). Multiplication by $\Phi(z)$ is an isomorphism from $\mathfrak{H}(\Sigma)$ onto $\mathfrak{H}(S)$. We denote this operator by U and show that

(4.4.4)
$$\begin{cases} T = U\big(T_\Sigma - F_\Sigma M G_\Sigma\big)U^*, \\ F = U F_\Sigma \Psi(0)^*, \end{cases}$$

where the operator $M \in \mathfrak{L}(\mathfrak{G}', \mathfrak{F}')$ is defined by

(4.4.5)
$$M = \begin{pmatrix} 0 & 0 \\ 0 & \Sigma_{22}(0)^{-1} \end{pmatrix} = \Psi(0)^* \begin{pmatrix} 0 & 0 \\ 0 & \sigma^{-1} \end{pmatrix}.$$

The first formula in (4.4.4) is equivalent to the assertion that for any $h \in \mathfrak{H}(\Sigma)$,

$$\frac{\Phi(z)h(z) - \Phi(0)h(0)}{z} = \Phi(z)\left[\frac{h(z) - h(0)}{z} - (F_\Sigma M G_\Sigma h)(z)\right].$$

This reduces to the identity

(4.4.6)
$$\Phi(z) - \Phi(0) = -\Phi(z)\big[\Sigma(z) - \Sigma(0)\big]M,$$

which is proved by matrix operations from the second relation in (4.3.10). The second formula in (4.4.4) is equivalent to the identity

$$\frac{S(z) - S(0)}{z} = \Phi(z)\frac{\Sigma(z) - \Sigma(0)}{z}\Psi(0)^*,$$

which holds by the second relation in (4.3.10). It follows that V is everywhere defined and continuous.

The rest is immediate from constructions in Chapter 2. By (4.4.2) a dense set in the graph of V^* is spanned by pairs of the form

$$\left(\begin{pmatrix} K_S(\alpha, \cdot)u_1 \\ u_2 \end{pmatrix}, \begin{pmatrix} \dfrac{K_S(\alpha, \cdot) - K_S(0, \cdot)}{\bar\alpha}u_1 + K_S(0, \cdot)u_2 \\ \dfrac{S(\alpha)^* - S(0)^*}{\bar\alpha}u_1 + S(0)^*u_2 \end{pmatrix}\right)$$

with $0 \neq \alpha \in \Omega(S)$. The proof of Theorem 2.2.1 then shows that V^* is an isometry. The colligation V is closely outer connected, because if $h \in \mathfrak{H}(S)$ and

$$GT^n h = 0, \qquad n \geq 0,$$

then the Taylor coefficients of h vanish, and hence h is the zero element of the space. The proof that $S(z) = \Theta_V(z)$ for all $z \in \Omega(S)\cap\Omega(\Theta_V)$ follows from (4.4.2) as in Theorem 2.2.1. $\qquad\square$

Alternative approach to Theorem 4.4.4. This follows similar lines, except we give a different proof that V is coisometric. First note that the two identities in (4.4.4) are part of a more general formula, namely,

$$(4.4.7) \qquad \begin{pmatrix} T & F \\ G & H \end{pmatrix} = \begin{pmatrix} U & 0 \\ 0 & \Phi(0) \end{pmatrix} \begin{pmatrix} T_\Sigma - F_\Sigma M G_\Sigma & F_\Sigma \\ G_\Sigma & \Delta \end{pmatrix} \begin{pmatrix} U^* & 0 \\ 0 & \Psi(0)^* \end{pmatrix},$$

where $\Delta \in \mathfrak{L}(\mathfrak{G}', \mathfrak{F}')$ is the operator

$$(4.4.8) \qquad \Delta = \begin{pmatrix} \Sigma_{11}(0) & 0 \\ 0 & -\Sigma_{22}(0) \end{pmatrix}$$

and the other notation is as above. We know that V_Σ is coisometric, that is,

$(\alpha) \qquad T_\Sigma T_\Sigma^* + F_\Sigma F_\Sigma^* = 1_{\mathfrak{H}(\Sigma)},$

$(\beta) \qquad T_\Sigma G_\Sigma^* + F_\Sigma H_\Sigma^* = 0,$

$(\gamma) \qquad G_\Sigma G_\Sigma^* + H_\Sigma H_\Sigma^* = 1_{\mathfrak{G}'}.$

These relations will be used to show that V is coisometric, that is,

$(\alpha') \qquad TT^* + FF^* = 1_{\mathfrak{H}(S)},$

$(\beta') \qquad TG^* + FH^* = 0,$

$(\gamma') \qquad GG^* + HH^* = 1_{\mathfrak{G}}.$

For brevity, write

$$(4.4.9) \qquad \Sigma(0) = \begin{pmatrix} \Sigma_{11} & \Sigma_{12} \\ \Sigma_{21} & \Sigma_{22} \end{pmatrix}.$$

Straightforward calculations using (4.3.6) and the definitions of $\Phi(0)$, $\Psi(0)$, Δ in (4.3.8), (4.3.9), (4.4.8) yield

$$(4.4.10) \qquad \begin{cases} \Phi(0) = \begin{pmatrix} 1_{\mathfrak{G}_+} & -\Sigma_{12}\Sigma_{22}^{-1} \\ 0 & -\tau^{-1}\Sigma_{22}^{-1} \end{pmatrix}, \\[2ex] \Psi(0) = \begin{pmatrix} 1_{\mathfrak{F}_+} & -\Sigma_{21}^*\Sigma_{22}^{*-1} \\ 0 & -\sigma^{-1}\Sigma_{22}^{*-1} \end{pmatrix}, \end{cases}$$

and

$$(4.4.11) \qquad \begin{cases} \Psi(0)^*\Psi(0) = \begin{pmatrix} 1_{\mathfrak{F}_+} & -\Sigma_{21}^*\Sigma_{22}^{*-1} \\ -\Sigma_{22}^{-1}\Sigma_{21} & \Sigma_{22}^{-1}\Sigma_{21}\Sigma_{21}^*\Sigma_{22}^{*-1} - \Sigma_{22}^{-1}\Sigma_{22}^{*-1} \end{pmatrix}, \\[2ex] \Psi(0)^*\Psi(0)\Delta^* = \begin{pmatrix} \Sigma_{11}^* & \Sigma_{21}^* \\ -\Sigma_{22}^{-1}\Sigma_{21}\Sigma_{11}^* & -\Sigma_{22}^{-1}\Sigma_{21}\Sigma_{21}^* + \Sigma_{22}^{-1} \end{pmatrix}, \\[2ex] \Delta\Psi(0)^*\Psi(0)\Delta^* = \begin{pmatrix} \Sigma_{11}\Sigma_{11}^* & \Sigma_{11}\Sigma_{21}^* \\ \Sigma_{21}\Sigma_{11}^* & \Sigma_{21}\Sigma_{21}^* - 1_{|\mathfrak{F}_-|} \end{pmatrix}. \end{cases}$$

Recall that $H_\Sigma = \Sigma(0)$. Thus by $(\alpha), (\beta), (\gamma)$, and (4.4.11),

$$TT^* + FF^*$$

$$= U\left[\left(T_\Sigma - F_\Sigma M G_\Sigma\right)\left(T_\Sigma^* - G_\Sigma^* M^* F_\Sigma^*\right) + F_\Sigma \Psi(0)^* \Psi(0) F_\Sigma^*\right]U^*$$

$$= U\left[1_{\mathfrak{H}(\Sigma)} - F_\Sigma F_\Sigma^* + F_\Sigma \Sigma(0)^* M^* F_\Sigma^* + F_\Sigma M \Sigma(0) F_\Sigma^*\right.$$

$$\left. + F_\Sigma M\left(1_{\mathfrak{G}'} - \Sigma(0)\Sigma(0)^*\right)M^* F_\Sigma^* + F_\Sigma \Psi(0)^* \Psi(0) F_\Sigma^*\right]U^*$$

$$= 1_{\mathfrak{H}(S)} + U F_\Sigma\left[-1_{\mathfrak{F}'} + \Sigma(0)^* M^* + M\Sigma(0)\right.$$

$$\left. + M\left(1_{\mathfrak{G}'} - \Sigma(0)\Sigma(0)^*\right)M^* + \Psi(0)^*\Psi(0)\right]F_\Sigma^* U^*$$

$$= 1_{\mathfrak{H}(S)} + U F_\Sigma\left[\begin{pmatrix} -1_{\mathfrak{F}_+} & 0 \\ 0 & -1_{|\mathfrak{G}_-|} \end{pmatrix} + \begin{pmatrix} 0 & \Sigma_{21}^* \Sigma_{22}^{*-1} \\ 0 & 1_{|\mathfrak{G}_-|} \end{pmatrix} + \begin{pmatrix} 0 & 0 \\ \Sigma_{22}^{-1}\Sigma_{21} & 1_{|\mathfrak{G}_-|} \end{pmatrix}\right.$$

$$\left. + \begin{pmatrix} 0 & 0 \\ 0 & \Sigma_{22}^{-1}\Sigma_{22}^{*-1} - \Sigma_{22}^{-1}\Sigma_{21}\Sigma_{21}^*\Sigma_{22}^{*-1} - 1_{|\mathfrak{G}_-|} \end{pmatrix} + \Psi(0)^*\Psi(0)\right]F_\Sigma^* U^*$$

$$= 1_{\mathfrak{H}(S)},$$

proving (α'). Similarly,

$$TG^* + FH^* = U\left[\left(T_\Sigma - F_\Sigma M G_\Sigma\right)G_\Sigma^* + F_\Sigma \Psi(0)^* \Psi(0)\Delta^*\right]\Phi(0)^*$$

$$= U\left[-F_\Sigma \Sigma(0)^* - F_\Sigma M\left(1_{\mathfrak{G}'} - \Sigma(0)\Sigma(0)^*\right)\right.$$

$$\left. + F_\Sigma \Psi(0)^* \Psi(0)\Delta^*\right]\Phi(0)^*$$

$$= U F_\Sigma\left[-\Sigma(0)^* - M\left(1_{\mathfrak{G}'} - \Sigma(0)\Sigma(0)^*\right) + \Psi(0)^*\Psi(0)\Delta^*\right]\Phi(0)^*$$

$$= U F_\Sigma\left[\begin{pmatrix} -\Sigma_{11}^* & -\Sigma_{21}^* \\ -\Sigma_{12}^* & -\Sigma_{22}^* \end{pmatrix} + \Psi(0)^*\Psi(0)\Delta^*\right.$$

$$\left. - \begin{pmatrix} 0 & 0 \\ -\Sigma_{22}^{-1}\Sigma_{21}\Sigma_{11}^* - \Sigma_{12}^* & \Sigma_{22}^{-1} - \Sigma_{22}^{-1}\Sigma_{21}\Sigma_{21}^* - \Sigma_{22}^* \end{pmatrix}\right]\Phi(0)^*$$

$$= 0,$$

The lower right 2×2 block,

$$(4.4.25) \qquad \begin{pmatrix} T_{22}(z) & 0 \\ 0 & S_{22}(z) \end{pmatrix} : \begin{pmatrix} -\mathbf{C} \\ \mathfrak{F}_- \end{pmatrix} \to \begin{pmatrix} -\mathbf{C} \\ \mathfrak{G}_- \end{pmatrix}$$

is not invertible at $z = 0$, but the lower right 3×3 block,

$$(4.4.26) \qquad \begin{pmatrix} T_{11}(z) & T_{12}(z) & 0 \\ T_{21}(z) & T_{22}(z) & 0 \\ 0 & 0 & S_{22}(z) \end{pmatrix} : \begin{pmatrix} \mathbf{C} \\ -\mathbf{C} \\ \mathfrak{F}_- \end{pmatrix} \to \begin{pmatrix} \mathbf{C} \\ -\mathbf{C} \\ \mathfrak{G}_- \end{pmatrix}$$

is invertible at $z = 0$.

Theorems 4.4.4, 4.4.5, and 4.4.6 cannot be applied with the fundamental decompositions

$$(4.4.27) \qquad \begin{cases} \hat{\mathfrak{F}} = (\mathfrak{F}_+ \oplus \mathbf{C}) \oplus (-\mathbf{C} \oplus \mathfrak{F}_-), \\ \hat{\mathfrak{G}} = (\mathfrak{G}_+ \oplus \mathbf{C}) \oplus (-\mathbf{C} \oplus \mathfrak{G}_-), \end{cases}$$

because of the noninvertibility of (4.4.25) when $z = 0$. The same conclusions can nevertheless be obtained by replacing the fundamental decompositions (4.2.27) by

$$(4.4.28) \qquad \begin{cases} \hat{\mathfrak{F}} = \mathfrak{F}_+ \oplus (\mathbf{M}^2 \oplus \mathfrak{F}_-), \\ \hat{\mathfrak{G}} = \mathfrak{G}_+ \oplus (\mathbf{M}^2 \oplus \mathfrak{G}_-). \end{cases}$$

The second summands, $\mathbf{M}^2 \oplus \mathfrak{F}_-$ and $\mathbf{M}^2 \oplus \mathfrak{G}_-$, are antispaces of Pontryagin spaces. The theory of the Potapov-Ginzburg transform in §4.3 can be extended to this case, however, and then the same methods used to prove Theorems 4.4.4, 4.4.5, and 4.4.6 apply.

4.5 Canonical models

Operators $T \in \mathfrak{L}(\mathfrak{H})$ and $R \in \mathfrak{L}(\mathfrak{K})$ on Pontryagin spaces $\mathfrak{H}$ and $\mathfrak{K}$ are called **unitarily equivalent** if there is an isomorphism $W \in \mathfrak{L}(\mathfrak{K}, \mathfrak{H})$ such that

$$R = W^{-1}TW.$$

In this situation, we sometimes view T as a model for R. It is natural to ask, which operators on Pontryagin spaces can be modeled by the main transformation T in some canonical coisometric, isometric, or unitary realization of a function $S(z)$ in $\mathbf{S}_\kappa(\mathfrak{F}, \mathfrak{G})$? We then call T a **canonical model** for R.

Recall from Theorem 1.3.4 that $\mathrm{ind}_- (1 - R^*R) = \mathrm{ind}_- (1 - RR^*)$ for any operator $R \in \mathfrak{L}(\mathfrak{K})$ whenever $\mathfrak{K}$ a Pontryagin space.

THEOREM 4.5.1. *Let $R \in \mathfrak{L}(\mathfrak{K})$, where $\mathfrak{K}$ is a Pontryagin space of negative index κ. Assume that*

(1) $\operatorname{ind}_- (1 - R^*R) = \nu < \infty$, *and*

(2) *the only closed subspace $\mathfrak{M}$ of $\mathfrak{K}$ such that*

$$R\mathfrak{M} \subseteq \mathfrak{M} \qquad and \qquad (1 - R^*R)\mathfrak{M} = \{0\}$$

is $\mathfrak{M} = \{0\}$.

Then R is unitarily equivalent to the main transformation $T \in \mathfrak{L}(\mathfrak{H}(S))$ in the canonical coisometric realization of a function $S(z)$ in $\mathbf{S}_\kappa(\mathfrak{F}, \mathfrak{G})$ for some Pontryagin spaces $\mathfrak{F}$ and $\mathfrak{G}$ such that $\operatorname{ind}_- \mathfrak{F} = \operatorname{ind}_- \mathfrak{G} = \nu$. The function $S(z)$ may be chosen so that the associated canonical coisometric colligation is a Julia colligation.

Proof. Define $S(z) = \Theta_{U_R}(z)$, where

$$(4.5.1) \qquad U_R = \begin{pmatrix} R & D \\ \tilde{D}^* & -L^* \end{pmatrix} \in \mathfrak{L}(\mathfrak{K} \oplus \mathfrak{D}, \mathfrak{K} \oplus \tilde{\mathfrak{D}})$$

is a Julia operator for R. By the hypothesis (2) and Theorem 1.3.2(5), U_R is closely outer connected. Take $\mathfrak{F} = \mathfrak{D}$ and $\mathfrak{G} = \tilde{\mathfrak{D}}$. By the hypothesis (1) and (1.3.8), $\mathfrak{F}$ and $\mathfrak{G}$ are Pontryagin spaces such that $\operatorname{ind}_- \mathfrak{F} = \operatorname{ind}_- \mathfrak{G} = \nu$. By Theorems 2.1.2(1) and 2.5.2, $S(z)$ belongs to $\mathbf{S}_\kappa(\mathfrak{F}, \mathfrak{G})$. Let

$$V = \begin{pmatrix} T & F \\ G & H \end{pmatrix} : \begin{pmatrix} \mathfrak{H}(S) \\ \mathfrak{F} \end{pmatrix} \to \begin{pmatrix} \mathfrak{H}(S) \\ \mathfrak{G} \end{pmatrix}$$

be the canonical coisometric colligation associated with $S(z)$. Since U_R and V are both coisometric, closely outer connected, and

$$\Theta_{U_R}(z) = S(z) = \Theta_V(z),$$

the colligations U_R and V are equivalent by Theorem 2.1.3(1). In particular, there is an isomorphism W from $\mathfrak{K}$ onto $\mathfrak{H}(S)$ such that $R = W^{-1}TW$. Since U_R is a Julia operator, so is V. $\qquad\square$

THEOREM 4.5.2. *Let $R \in \mathfrak{L}(\mathfrak{K})$, where $\mathfrak{K}$ is a Pontryagin space of negative index κ. Assume that*

(1) $\operatorname{ind}_- (1 - RR^*) = \nu < \infty$, *and*

(2) *the only closed subspace $\mathfrak{M}$ of $\mathfrak{K}$ such that*

$$R^*\mathfrak{M} \subseteq \mathfrak{M} \qquad and \qquad (1 - RR^*)\mathfrak{M} = \{0\}$$

is $\mathfrak{M} = \{0\}$.

Then R is unitarily equivalent to the main transformation $\tilde{T} \in \mathfrak{L}(\mathfrak{H}(\tilde{S}))$ in the canonical isometric realization of a function $S(z)$ in $\mathbf{S}_\kappa(\mathfrak{F}, \mathfrak{G})$ for some Pontryagin spaces $\mathfrak{F}$ and $\mathfrak{G}$ such that $\operatorname{ind}_- \mathfrak{F} = \operatorname{ind}_- \mathfrak{G} = \nu$. The function $S(z)$ may be chosen so that the associated canonical isometric colligation is a Julia colligation.

THEOREM 4.5.3. *Let $R \in \mathfrak{L}(\mathfrak{K})$, where $\mathfrak{K}$ is a Pontryagin space of negative index κ. Assume that*

(1) $\mathrm{ind}_-\,(1 - R^*R) = \nu < \infty$, *and*

(2) *the only closed subspace $\mathfrak{M}$ of $\mathfrak{K}$ such that*

$$R\,\mathfrak{M} = \mathfrak{M} \qquad and \qquad (1 - R^*R)\,\mathfrak{M} = \{0\}$$

is $\mathfrak{M} = \{0\}$.

Then R is unitarily equivalent to the main transformation $A \in \mathfrak{L}(\mathfrak{D}(S))$ in the canonical unitary realization of a function $S(z)$ in $\mathbf{S}_\kappa(\mathfrak{F}, \mathfrak{G})$ for some Pontryagin spaces $\mathfrak{F}$ and $\mathfrak{G}$ such that $\mathrm{ind}_-\,\mathfrak{F} = \mathrm{ind}_-\,\mathfrak{G} = \nu$. The function $S(z)$ may be chosen so that the associated canonical unitary colligation is a Julia colligation.

Proofs of Theorems 4.5.2 and 4.5.3. As these arguments directly parallel the proof of Theorem 4.5.1, a sketch of the details will be sufficient.

Define $S(z) = \Theta_{U_R}(z)$ as in the proof of Theorem 4.5.1. For Theorem 4.5.2 we use Theorem 1.3.2(4) to conclude that U_R is closely inner connected. For Theorem 4.5.3, Theorem 1.3.2(6) shows that U_R is closely connected. In both cases take $\mathfrak{F} = \mathfrak{D}$ and $\mathfrak{G} = \tilde{\mathfrak{D}}$ and use (1.3.8) to obtain $\mathrm{ind}_-\,\mathfrak{F} = \mathrm{ind}_-\,\mathfrak{G} = \nu$. By Theorem 2.1.2(2) or 2.1.2(3), $S(z)$ belongs to $\mathbf{S}_\kappa(\mathfrak{F}, \mathfrak{G})$. By Theorem 2.1.3(2) or 2.1.3(3), U_R is equivalent to the canonical isometric colligation or canonical unitary colligation associated with $S(z)$, yielding the results.   □

A special case of Theorem 4.5.3 is a completely nonunitary contraction $R \in \mathfrak{L}(\mathfrak{K})$, where $\mathfrak{K}$ is a Hilbert space ("completely nonunitary" means that R has no nontrivial reducing subspace on which it is unitary). Following Sz.-Nagy and Foiaş [1970], p. 237, write

$$D_R = (1 - R^*R)^{1/2}, \qquad \mathfrak{D}_R = \overline{\mathrm{ran}}\,D_R\,,$$

$$D_{R^*} = (1 - RR^*)^{1/2}, \qquad \mathfrak{D}_{R^*} = \overline{\mathrm{ran}}\,D_{R^*}\,.$$

Then in Theorem 4.5.3 we can choose the operator-valued function with values in $\mathfrak{L}(\mathfrak{D}_{R^*}, \mathfrak{D}_R)$ given by

$$S(z) = \left[-R^* + zD_R(1 - zR)^{-1}D_{R^*} \right]\Big|_{\mathfrak{D}_{R^*}}\,.$$

This is the characteristic function for the contraction operator R^* as written by Sz.-Nagy and Foiaş. It is obtained from the proof of Theorem 4.5.3 by choosing

$\mathfrak{D} = \mathfrak{D}_{R^*}$, $\tilde{\mathfrak{D}} = \mathfrak{D}_R$, and

$$\tilde{D} = (1 - R^*R)^{1/2}|_{\mathfrak{D}_{R^*}},$$

$$D = (1 - RR^*)^{1/2}|_{\mathfrak{D}_R},$$

$$L = R|_{\mathfrak{D}_{R^*}},$$

in (4.5.1). The properties of a Julia operator are verified by standard calculations with these choices.

In Theorems 4.5.1–4.5.3, the function $S(z)$ is essentially unique in a natural sense as described in the next result.

THEOREM 4.5.4. *For each* $j = 1, 2$, *let* $\mathfrak{H}_j, \mathfrak{F}_j, \mathfrak{G}_j$ *be Pontryagin spaces, and let*

(4.5.2)
$$\begin{pmatrix} T_j & F_j \\ G_j & H_j \end{pmatrix} : \begin{pmatrix} \mathfrak{H}_j \\ \mathfrak{F}_j \end{pmatrix} \to \begin{pmatrix} \mathfrak{H}_j \\ \mathfrak{G}_j \end{pmatrix}$$

be Julia colligations with characteristic functions

$$S_j(z) = H_j + zG_j(1 - zT_j)^{-1}F_j.$$

If T_1 *and* T_2 *are unitarily equivalent, there exist isomorphisms* $\varphi : \mathfrak{F}_1 \to \mathfrak{F}_2$ *and* $\psi : \mathfrak{G}_1 \to \mathfrak{G}_2$ *such that* $S_1(z) \equiv \psi^{-1}S_2(z)\varphi$.

Proof. Since T_1 and T_2 are unitarily equivalent, there exists an isomorphism $U : \mathfrak{H}_1 \to \mathfrak{H}_2$ such that $T_1 = U^{-1}T_2U$. One Julia operator for T_1 is given by (4.5.2) with $j = 1$, and another is

$$\begin{pmatrix} U^{-1} & 0 \\ 0 & 1_{\mathfrak{G}_2} \end{pmatrix} \begin{pmatrix} T_2 & F_2 \\ G_2 & H_2 \end{pmatrix} \begin{pmatrix} U & 0 \\ 0 & 1_{\mathfrak{F}_2} \end{pmatrix} = \begin{pmatrix} T_1 & U^{-1}F_2 \\ G_2U & H_2 \end{pmatrix} : \begin{pmatrix} \mathfrak{H}_1 \\ \mathfrak{F}_2 \end{pmatrix} \to \begin{pmatrix} \mathfrak{H}_1 \\ \mathfrak{G}_2 \end{pmatrix}.$$

By (1.3.8) and our assumption that $\mathfrak{F}_1$ and $\mathfrak{G}_1$ are Pontryagin spaces, T_1 satisfies (1.3.13). Hence by Theorem 1.3.3, T_1 has an essentially unique Julia operator. This means that there are isomorphisms $\varphi : \mathfrak{F}_1 \to \mathfrak{F}_2$ and $\psi : \mathfrak{G}_1 \to \mathfrak{G}_2$ such that

$$F_1 = U^{-1}F_2\varphi, \quad G_1^* = (G_2U)^*\psi, \quad \varphi H_1^* = H_2^*\psi,$$

by (1.3.9) and (1.3.10). So

$$\begin{aligned} S_1(z) &= H_1 + zG_1(1 - zT_1)^{-1}F_1 \\ &= \psi^{-1}H_2\varphi + z\psi^{-1}G_2U(1 - zT_1)^{-1}U^{-1}F_2\varphi \\ &= \psi^{-1}\left[H_2 + zG_2(1 - zT_2)^{-1}F_2\right]\varphi \\ &= \psi^{-1}S_2(z)\varphi \end{aligned}$$

identically, as was to be shown.   □

Theorems 4.5.1, 4.5.2, and 4.5.3 do not cover all possibilities, but some restrictions are needed to draw the same conclusions. Notice, for example, that Theorem 4.2.3 gives a necessary condition in order that an operator $R \in \mathfrak{L}(\mathfrak{K})$, with $\mathfrak{K}$ a Pontryagin space, be unitarily equivalent to the main transformation in $\mathfrak{H}(S)$ for some $S(z)$ in $\mathbf{S}_\kappa(\mathfrak{F}, \mathfrak{G})$ with $\mathfrak{F}$ and $\mathfrak{G}$ Hilbert spaces. Namely, both R and R^* must have unique maximal nonpositive invariant subspaces, and these subspaces must be uniformly negative.

In some sense, it is possible to completely characterize the transformation $T : [h(z) - h(0)]/z$ in a space $\mathfrak{H}(S)$ for functions $S(z)$ which belong to some class $\mathbf{S}_\kappa(\mathfrak{F}, \mathfrak{G})$ with $\mathrm{ind}_- \mathfrak{F} = \mathrm{ind}_- \mathfrak{G} < \infty$. The characterization depends on finding another operator having certain properties, however, and is thus not intrinsic.

THEOREM 4.5.5. *Let $R \in \mathfrak{L}(\mathfrak{K})$, where $\mathfrak{K}$ is a Pontryagin space of negative index κ. The following assertions are equivalent:*

(1) *R is unitarily equivalent to the main transformation T in the canonical coisometric realization of a function $S(z)$ in some class $\mathbf{S}_\kappa(\mathfrak{F}, \mathfrak{G})$ where $\mathfrak{F}$ and $\mathfrak{G}$ are Pontryagin spaces having the same negative index;*

(2) *there exist a Pontryagin space $\mathfrak{A}$ and an operator $A \in \mathfrak{L}(\mathfrak{K}, \mathfrak{A})$ such that $1 - R^*R \geq A^*A$ and the only $k \in \mathfrak{K}$ such that $AR^n k = 0$ for all $n \geq 0$ is $k = 0$.*

In this case, $S(z)$ in $\mathbf{S}_\kappa(\mathfrak{F}, \mathfrak{G})$ can be chosen in (1) such that $\mathrm{ind}_- \mathfrak{F} = \mathrm{ind}_- \mathfrak{G} = \nu$ where $\nu = \mathrm{ind}_- \mathfrak{A}$ for any space $\mathfrak{A}$ as in (2).

Proof. (1) $\Rightarrow$ (2) Assume (1). Without loss of generality, we can take $R = T$. Choose $\mathfrak{A} = \mathfrak{G}$, and let $A = G$ be evaluation at the origin. Then $1 - T^*T \geq G^*G$ by the last assertion in Theorem 3.1.2. The condition $GT^n h = 0$, $n \geq 0$, asserts that the Taylor coefficients of h vanish, and therefore $h \equiv 0$.

(2) $\Rightarrow$ (1) Assume (2). Define $\mathfrak{G} = \mathfrak{A}$, and let $\mathfrak{P}$ be the linear space of holomorphic functions with values in $\mathfrak{G}$ of the form

$$h(z) = A(1 - zR)^{-1}k$$

for some $k \in \mathfrak{K}$. The domain of the functions in $\mathfrak{P}$ is taken to be a fixed subregion of the unit disk containing the origin for which the expressions are defined. The hypotheses in (2) imply that the mapping $W : k \to h(z)$ which is defined in this way is one-to-one. We give $\mathfrak{P}$ the inner product which makes W an isomorphism from $\mathfrak{K}$ onto $\mathfrak{P}$. Then $\mathfrak{P}$ is a Pontryagin space of negative index κ which has a reproducing kernel. From

$$W : k \to h(z)$$

we obtain

$$W : Rk \to [h(z) - h(0)]/z.$$

The assumption $1 - R^*R \geq A^*A$ implies that (3.1.1) holds. By Theorem 3.1.2, $\mathfrak{P}$ is isometrically equal to $\mathfrak{H}(S)$ for some function $S(z)$ as in (1). Thus R is unitarily equivalent to the main transformation T in $\mathfrak{H}(S)$, and this proves (1).

By the preceding construction, $\mathrm{ind}_- \, \mathfrak{F} = \mathrm{ind}_- \, \mathfrak{G} = \nu$ where $\nu = \mathrm{ind}_- \, \mathfrak{A}$ for any space $\mathfrak{A}$ as in (2). $\square$

EXAMPLE 1. *Any contraction R on a finite-dimensional Pontryagin space $\mathfrak{K}$ satisfies the conditions in Theorem 4.5.5(2) and hence has a canonical model as in Theorem 4.5.5(1).*

To see this, let $\mathfrak{A}$ be any Pontryagin space such that
$$\mathrm{ind}_\pm \, \mathfrak{A} \geq \max \left\{ \mathrm{ind}_\pm \, \mathfrak{K} \right\}.$$
This allows us to choose an operator $A \in \mathfrak{L}(\mathfrak{K}, \mathfrak{A})$ such that $\ker A = \{0\}$ and $A^*A = 0$ (the latter condition is assured by choosing A so that its range lies in a neutral subspace of $\mathfrak{A}$). Thus $1 - R^*R \geq 0 = A^*A$ and $AR^nk = 0$ for all $n \geq 0$ only for $k = 0$.

In the following special case, a choice of $S(z)$ is exhibited.

EXAMPLE 2. *Let $R = 1_{\mathfrak{K}}$ be the identity operator on a finite-dimensional Pontryagin space $\mathfrak{K}$. Choose a Pontryagin space $\mathfrak{G}$ and $A \in \mathfrak{L}(\mathfrak{K}, \mathfrak{G})$ such that $\ker A = \{0\}$ and $A^*A = 0$. Put $N = AA^* \in \mathfrak{L}(\mathfrak{G})$. If $\mathfrak{K}$ has negative index κ, the function*
$$S(z) = 1_{\mathfrak{G}} + \frac{z+1}{z-1} \, N$$
is in $\mathbf{S}_\kappa(\mathfrak{G})$, and R is unitarily equivalent to $T : h(z) \to [h(z) - h(0)]/z$ in $\mathfrak{H}(S)$.

The situation here is different from Theorem 4.5.1. There we were able to take $S(z) = \Theta_{U_R}(z)$ for any Julia operator U_R for R. In Example 2, this prescription produces $S(z) \equiv 0$ on a zero-dimensional space, which clearly will not do. To see that the prescribed choice of $S(z)$ works, notice that $N^2 = 0$. Hence a short calculation yields
$$K_S(w, z) = \frac{2AA^*}{(1-z)(1-\bar{w})}.$$
This is the reproducing kernel for the Pontryagin space of functions of the form
$$h(z) = \frac{\sqrt{2}A}{1-z} \, k, \qquad k \in \mathfrak{K},$$
in the inner product such that
$$W : k \to \sqrt{2}Ak/(1-z)$$
is an isomorphism. Hence $S(z)$ is in $\mathbf{S}_\kappa(\mathfrak{G})$. From the explicit description of $\mathfrak{H}(S)$ we see that T acts like the identity on every element of $\mathfrak{H}(S)$, and hence $R = W^{-1}TW$.

we obtain

$$\hat{K}_S(w,z) = \hat{A}(\ell)\,\frac{N(\ell) - N(\lambda)^*}{\ell - \bar{\lambda}}\,\hat{A}(\lambda)^*$$

whenever $z = z(\ell)$, $w = z(\lambda)$ for points ℓ, λ in $\mathbf{C}_+ \cup \mathbf{C}_-$. Therefore multiplication by $\hat{A}(\ell)$ is an isomorphism which maps the Pontryagin space with reproducing kernel (E4) on $(\mathbf{C}_+ \cup \mathbf{C}_-) \times (\mathbf{C}_+ \cup \mathbf{C}_-)$ onto $\hat{\mathfrak{D}}(S)$.

More generally, similar constructions can be made with (E4) replaced by kernels of the form

(E5)
$$\frac{N(\ell)M(\lambda)^* - M(\ell)N(\lambda)^*}{\ell - \bar{\lambda}}$$

for pairs $M(\ell)$ and $N(\ell)$ of operator-valued functions meeting appropriate conditions that extend the notion of a generalized Nevanlinna function.

The canonical coisometric, isometric, and unitary realizations of generalized Schur functions $S(z)$ lead to operator representations of generalized Nevanlinna functions $N(\ell)$ in the case of kernels (E4) and to representations of pairs $M(\ell)$ and $N(\ell)$ for kernels (E5).

APPENDIX:
SOME FINITE-DIMENSIONAL SPACES

We assume that the coefficient spaces $\mathfrak{F}$ and $\mathfrak{G}$ are Hilbert spaces, and in this case we characterize all finite-dimensional spaces $\mathfrak{H}(S)$ which are Hilbert spaces, or antispaces of Hilbert spaces, for which the associated canonical coisometric colligation is unitary. These results are used in the derivation of the Kreĭn-Langer factorization in §4.2. Some methods from §4.1 are used in the constructions, notably Theorem 4.1.1.

By a **Blaschke-Potapov factor** we mean a function of the form

$$(A1) \qquad \qquad B(z) = 1 - P + \rho \, \frac{z - \alpha}{1 - \bar{\alpha}z} \, P,$$

where $P \in \mathfrak{L}(\mathfrak{G})$ is a nonzero projection, $|\alpha| < 1$, and $|\rho| = 1$. The factor is called **simple** if P has rank 1. For such a factor, $B(0)$ is invertible if and only if $\alpha \neq 0$. We first give some one-dimensional examples which show the role of unitarity in the canonical coisometric realization.

If $\mathfrak{G}$ is a Hilbert space, the **Hardy class** $H_{\mathfrak{G}}^2$ is the Hilbert space of holomorphic functions

$$h(z) = h_0 + h_1 z + h_2 z^2 + \cdots$$

on the unit disk with values in $\mathfrak{G}$ such that $\|h\|^2 = \sum_{n=0}^{\infty} \|h_n\|_{\mathfrak{G}}^2 < \infty$.

EXAMPLE 1. *Let $\mathfrak{G}$ be a Hilbert space, and let $B(z)$ be a simple Blaschke-Potapov factor (A1) with values in $\mathfrak{L}(\mathfrak{G})$.*

(1) *If $S(z) = B(z)$, then $S(z)$ is in $\mathbf{S}_0(\mathfrak{G})$ and $\mathfrak{H}(S)$ is a one-dimensional Hilbert space which is contained isometrically in $H_{\mathfrak{G}}^2$.*

(2) *If $B(0)$ is invertible and $S(z) = B(z)^{-1}$, then $S(z)$ is in $\mathbf{S}_1(\mathfrak{G})$ and $\mathfrak{H}(S)$ is the antispace of a one-dimensional Hilbert space.*

In both cases, the canonical coisometric and isometric colligations associated with $S(z)$ are unitary.

The two cases can be combined by writing $\gamma = \alpha$ in case (1) and $\gamma = 1/\bar{\alpha}$ in case (2). Then in both cases,

$$K_S(w, z) = \frac{1 - |\gamma|^2}{(1 - \bar{\gamma}z)(1 - \gamma\bar{w})} \, P.$$

In case (1), if g_0 is a unit vector in the range of P,

$$h_0(z) = K_S(0, z) \frac{g_0}{\sqrt{1 - |\alpha|^2}} = \sqrt{1 - |\alpha|^2} \frac{g_0}{1 - \bar{\alpha} z}$$

is an element of $\mathfrak{H}(S)$ which spans the space and satisfies $\langle h_0, h_0 \rangle_{\mathfrak{H}(S)} = 1$. Also $\langle h_0, h_0 \rangle_{H^2_{\mathfrak{G}}} = 1$. In case (2), if g_0 is a unit vector in the range of P,

$$h_0(z) = K_S(0, z) \frac{\bar{\alpha} \, g_0}{\sqrt{1 - |\alpha|^2}} = \sqrt{1 - |\alpha|^2} \frac{g_0}{z - \alpha}$$

is an element of $\mathfrak{H}(S)$ which spans the space and satisfies $\langle h_0, h_0 \rangle_{\mathfrak{H}(S)} = -1$. Clearly $S(z)$ belongs to $\mathbf{S}_0(\mathfrak{G})$ in case (1) and to $\mathbf{S}_1(\mathfrak{G})$ in case (2).

For both parts (1) and (2), we can verify the unitarity of the coisometric and isometric colligations using the criteria in Theorems 3.2.3(2) and 3.3.3(2). For this purpose, notice that $\tilde{S}(z)$ has the same form as $S(z)$ but with ρ, α replaced by $\bar{\rho}, \bar{\alpha}$, and hence the form of $\mathfrak{H}(\tilde{S})$ is also known from what is shown above.

EXAMPLE 2. *Let $\mathfrak{E}, \mathfrak{F}, \mathfrak{G}$ be Hilbert spaces with $\mathfrak{F} = \mathfrak{G} \oplus \mathfrak{E}$ and $\dim \mathfrak{E} = 1$. Let $E \in \mathfrak{L}(\mathfrak{E}, \mathfrak{G})$ be an isometry, and let $P = EE^* \in \mathfrak{L}(\mathfrak{G})$ be the projection onto its range. Define a function with values in $\mathfrak{L}(\mathfrak{F}, \mathfrak{G})$ by*

$$\text{(A2)} \qquad\qquad S(z) = \left(1 - P + \rho \, \frac{z - \gamma}{1 - \bar{\gamma} z} \, P \quad \sigma E \right),$$

where γ, ρ, σ are complex numbers such that $|\gamma| \neq 1$, $|\rho|^2 + |\sigma|^2 = 1$, and $\rho\sigma \neq 0$.

(1) *If $|\gamma| < 1$, then $S(z)$ belongs to $\mathbf{S}_0(\mathfrak{F}, \mathfrak{G})$, $\mathfrak{H}(S)$ is a one-dimensional Hilbert space which is contained contractively but not isometrically in $H^2_{\mathfrak{G}}$, and $\mathfrak{H}(\tilde{S})$ is infinite-dimensional and contained isometrically in $H^2_{\mathfrak{F}}$.*

(2) *If $|\gamma| > 1$, then $S(z)$ belongs to $\mathbf{S}_1(\mathfrak{F}, \mathfrak{G})$, $\mathfrak{H}(S)$ is the antispace of a one-dimensional Hilbert space, and $\mathfrak{H}(\tilde{S})$ is an infinite-dimensional Pontryagin space having negative index 1.*

In both cases, $S(z)f \equiv 0$ for some $f \in \mathfrak{F}$ only for $f = 0$. Also in both cases, the canonical coisometric colligation is not unitary, and the canonical isometric colligation is unitary.

Consider case (1). A short calculation using the formula (A2) for $S(z)$ yields

$$K_S(w, z) = \frac{|\rho|^2 (1 - |\gamma|^2)}{(1 - \bar{\gamma} z)(1 - \gamma \bar{w})} \, P.$$

Since $|\gamma| < 1$, $S(z)$ belongs to $\mathbf{S}_0(\mathfrak{F}, \mathfrak{G})$ and $\mathfrak{H}(S)$ is a one-dimensional Hilbert space which is contained contractively but not isometrically in $H^2_{\mathfrak{G}}$. To see that the space $\mathfrak{H}(\tilde{S})$ is contained isometrically in $H_{\tilde{\mathfrak{F}}}^2$, notice that multiplication by $\tilde{S}(z)$ acts as an isometry from $H_{\mathfrak{G}}^2$ into $H_{\tilde{\mathfrak{F}}}^2$. Its range $\tilde{S}H_{\mathfrak{G}}^2$, considered in the inner product of $H_{\tilde{\mathfrak{F}}}^2$, has the reproducing kernel $\tilde{S}(z)\tilde{S}(w)^*/(1-z\bar{w})$ by Theorem 1.5.7. Since

$$K_{\tilde{S}}(w,z) = \frac{1_{\tilde{\mathfrak{F}}}}{1-z\bar{w}} - \frac{\tilde{S}(z)\tilde{S}(w)^*}{1-z\bar{w}},$$

the space $\mathfrak{H}(\tilde{S})$ coincides with the orthogonal complement of $\tilde{S}H_{\mathfrak{G}}^2$ in $H_{\tilde{\mathfrak{F}}}^2$ in the inner product of $H_{\tilde{\mathfrak{F}}}^2$, by standard properties of reproducing kernel Hilbert spaces (see Theorem 1.5.6). Explicitly,

$$K_{\tilde{S}}(w,z) = \frac{1}{1-z\bar{w}} \begin{pmatrix} \left[1 - |\rho|^2 \dfrac{z-\bar{\gamma}}{1-\gamma z}\dfrac{\bar{w}-\gamma}{1-\bar{\gamma}\bar{w}}\right]P & -\bar{\rho}\sigma\dfrac{z-\bar{\gamma}}{1-\gamma z}E \\[4mm] -\rho\bar{\sigma}\dfrac{\bar{w}-\gamma}{1-\bar{\gamma}\bar{w}}E^* & |\rho|^2 \cdot 1_{\mathfrak{E}} \end{pmatrix}.$$

Positivity of $|\rho|^2$ in the 22-entry of the matrix assures that there are Gram matrices of elements of $\mathfrak{H}(\tilde{S})$ of arbitrarily large size which are nonnegative and invertible. It follows that $\mathfrak{H}(\tilde{S})$ is infinite-dimensional.

In case (2), the formulas given above for $K_S(w,z)$ and $K_{\tilde{S}}(w,z)$ still hold. Since now $|\gamma| > 1$, $\mathfrak{H}(S)$ is the antispace of a one-dimensional Hilbert space. Hence $S(z)$ is in $\mathbf{S}_1(\mathfrak{F},\mathfrak{G})$ by Theorem 2.5.2. Therefore $\mathfrak{H}(\tilde{S})$ is a Pontryagin space of negative index 1. As in part (1), the formula for $K_{\tilde{S}}(w,z)$ may be used to show that there are arbitrarily large nonnegative and invertible Gram matrices in $\mathfrak{H}(\tilde{S})$, so $\mathfrak{H}(\tilde{S})$ is infinite-dimensional.

In both parts (1) and (2), the identity $S(z)f \equiv 0$ is possible only for $f = 0$ by the definition (A2) of $S(z)$. Let e_0 be a unit vector in $\mathfrak{E}$. Then $g_0 = Ee_0$ is a unit vector in the range of P. In both parts (1) and (2), the function

$$S(z)\begin{pmatrix} \bar{\gamma}\sigma g_0 \\ \rho e_0 \end{pmatrix} = (1-P)\bar{\gamma}\sigma g_0 + \rho\frac{z-\gamma}{1-\bar{\gamma}z}P\bar{\gamma}\sigma g_0 + \sigma\rho g_0 = \frac{\rho\sigma(1-|\gamma|^2)g_0}{1-\bar{\gamma}z}$$

belongs to $\mathfrak{H}(S)$, and hence the canonical coisometric colligation associated with $S(z)$ is not unitary by Theorem 3.2.3(2). It remains to show that in both cases (1) and (2), the canonical isometric colligation associated with $S(z)$ is unitary. We do this by verifying the condition in Theorem 3.3.5(4): for all $g_1, g_2 \in \mathfrak{G}$,

$$\left\langle \frac{\tilde{S}(z)-\tilde{S}(0)}{z}g_1, \frac{\tilde{S}(z)-\tilde{S}(0)}{z}g_2 \right\rangle_{\mathfrak{H}(\tilde{S})} = \langle K_S(0,0)g_1, g_2\rangle_{\mathfrak{G}}.$$

Since

$$\frac{\tilde{S}(z)-\tilde{S}(0)}{z} = \bar{\rho}\left(1-|\gamma|^2\right)\begin{pmatrix} P/(1-\gamma z) \\ 0 \end{pmatrix} = K_{\tilde{S}}(0,z)\begin{pmatrix} \bar{\rho}P \\ -\bar{\sigma}\gamma E^* \end{pmatrix},$$

the identity to be verified is equivalent to

$$(\rho P \quad -\sigma \bar{\gamma} E) \, K_{\tilde{S}}(0,0) \begin{pmatrix} \bar{\rho} P \\ -\bar{\sigma} \gamma E^* \end{pmatrix} = K_S(0,0).$$

We easily check that both sides in the last formula reduce to $|\rho|^2 \left(1 - |\gamma|^2 \right) P$, and thus all of the assertions in Example 2 are verified.

In some sense, Examples 1–2 account for all one-dimensional spaces $\mathfrak{H}(S)$.

THEOREM A1. *Let $\mathfrak{G}$ be a Hilbert space. Suppose that $\mathfrak{P}$ is either a one-dimensional Hilbert space or the antispace of a one-dimensional Hilbert space of functions with values in $\mathfrak{G}$ satisfying the hypotheses of Theorem 3.1.2. Then $\mathfrak{P}$ is isometrically equal to $\mathfrak{H}(S)$ where $S(z)$ is one of the types in Examples 1–2.*

Proof. By Theorem 3.1.2, $\mathfrak{P}$ has the form $\mathfrak{H}(S)$, where $S(z)$ is in $\mathbf{S}_\kappa(\mathfrak{F}, \mathfrak{G})$, $\kappa = 0$ or 1, for some Hilbert space $\mathfrak{F}$. If we require that $S(z)f \equiv 0$ only for $f = 0$, then the choice of $S(z)$ is unique up to a constant right unitary factor by Theorem 3.1.3. In both cases, let T be the transformation $h(z)$ into $[h(z) - h(0)]/z$ in $\mathfrak{P}$.

First suppose $\kappa = 0$. Then $\mathfrak{P}$ is a one-dimensional Hilbert space and contains an element h_0 of norm one such that $Th_0 = \bar{\gamma} h_0$ for some complex number γ. Thus

$$[h_0(z) - h_0(0)]/z = \bar{\gamma} h_0(z),$$

and so $h_0(z) = c_0 g_0/(1 - \bar{\gamma} z)$, where c_0 is a nonzero complex number and g_0 is a unit vector in $\mathfrak{G}$. By (3.1.1) and the choice of h_0,

$$|\gamma|^2 \leq 1 - |c_0|^2.$$

In particular, $|\gamma| < 1$. By adjusting c_0 by a constant factor of unit modulus, we may choose

$$h_0(z) = |\rho| \frac{\sqrt{1 - |\gamma|^2}}{1 - \bar{\gamma} z} g_0,$$

where ρ is a nonzero complex number such that $|\rho| \leq 1$. The reproducing kernel for $\mathfrak{P}$ is given by

$$K_{\mathfrak{P}}(w, z) = \frac{|\rho|^2 (1 - |\gamma|^2)}{(1 - \bar{\gamma} z)(1 - \gamma \bar{w})} P,$$

where P is the projection onto the span of g_0. We can thus choose $S(z)$ as in Example 1(1) with $\alpha = \gamma$ if $|\rho| = 1$ and as in Example 2(1) if $|\rho| < 1$.

When $\kappa = 1$, $\mathfrak{P}$ is the antispace of a one-dimensional Hilbert space. Then $Th_0 = \bar{\gamma} h_0$ for some complex number γ and some element h_0 of $\mathfrak{P}$ such that

$\langle h_0, h_0 \rangle_{\mathfrak{P}} = -1$. As before, $h_0(z) = c_0 g_0 / (1 - \bar{\gamma} z)$, where c_0 is a nonzero complex number and g_0 is a unit vector in $\mathfrak{G}$. Now (3.1.1) implies

$$-|\gamma|^2 \leq -1 - |c_0|^2.$$

So $|\gamma| > 1$, and an adjustment of c_0 by a constant factor of unit modulus yields

$$h_0(z) = |\rho| \frac{\sqrt{|\gamma|^2 - 1}}{1 - \bar{\gamma} z} g_0,$$

where $0 < |\rho| \leq 1$. The reproducing kernel for $\mathfrak{P}$ has the same form as before, except now $|\gamma| > 1$. In this case, we can choose $S(z)$ as in Example 1(2) with $\alpha = 1/\bar{\gamma}$ if $|\rho| = 1$ and as in Example 2(2) if $|\rho| < 1$.   □

THEOREM A2. *Let $\mathfrak{G}$ be a Hilbert space, and let $B(z)$ be a product of n simple Blaschke-Potapov factors and a constant unitary operator in $\mathfrak{L}(\mathfrak{G})$. Then*

(1) *$B(z)$ is in $\mathbf{S}_0(\mathfrak{G})$ and* dim $\mathfrak{H}(B) = n$, *and*
(2) *the canonical coisometric colligation V associated with $B(z)$ is unitary.*

Conversely, every function $B(z)$ satisfying (1) and (2) has this form. For any such function $B(z)$, $\mathfrak{H}(B)$ is contained isometrically in the Hardy class $H^2_{\mathfrak{G}}$

We call such a function $B(z)$ a **Blaschke product of degree** n. Degree is well defined: although $B(z)$ can in general be written in many ways as a finite product of simple Blaschke-Potapov factors and a constant unitary factor, the number of factors is equal to the dimension of $\mathfrak{H}(B)$ by Theorem A2. The constant unitary factor can be written on either side. A Blaschke product of degree 0 is a function identically equal to a unitary operator $U \in \mathfrak{L}(\mathfrak{G})$.

The values of a Blaschke product $B(z)$ of degree n are unitary on the unit circle $|z| = 1$. In the unit disk $|z| < 1$, the values of $B(z)$ are invertible except at certain isolated points. In any fixed representation of $B(z)$ involving n simple Blaschke-Potapov factors, the exceptional values coincide with the numbers α in those factors (A1) that appear in the representation. The distinct points obtained in this way, say $\alpha_1, \ldots, \alpha_k$, are thus independent of the representation. They are called the **null points** of $B(z)$. One can verify that the null points of $B(z)$ coincide with the points w in the unit disk such that the equation $B(w)^* g = 0$ has a nonzero solution g in $\mathfrak{G}$, or, equivalently, with the points w in the unit disk such that the equation $B(w)g = 0$ has a nonzero solution g in $\mathfrak{G}$.

In the situation of the theorem, the null points of $B(z)$ can be characterized in terms of the main transformation T in the canonical coisometric colligation, that is, the transformation $h(z) \to [h(z) - h(0)]/z$ in $\mathfrak{H}(B)$. Namely, the spectrum of T is given by

$$\sigma(T) = \{ \bar{\alpha}_1, \ldots, \bar{\alpha}_k \},$$

where $\alpha_1, \ldots, \alpha_k$ are the null points of $B(z)$. In fact, the proof of the theorem will show that every eigenvalue of T is the conjugate of a null point of $B(z)$. On the other hand, if α is a null point of $B(z)$, there is a nonzero vector g in $\mathfrak{G}$ such that $B(\alpha)^*g = 0$. Then

$$\frac{g}{1 - \bar{\alpha}z} = K_B(\alpha, z)g$$

belongs to $\mathfrak{H}(B)$, and so $\bar{\alpha} \in \sigma(T)$.

Proof of Theorem A2. The case $n = 1$ in the direct statement is clear (see Example 1).

We proceed by induction, assuming that (1) and (2) have been verified for products of n factors (A1) and a constant unitary operator. At the next stage, we can write $B(z) = B_0(z)B_n(z)$, where

(A3) $$B_0(z) = 1 - P_0 + \rho_0 \frac{z - \alpha_0}{1 - \bar{\alpha}_0 z} P_0$$

is a simple Blaschke-Potapov factor and $B_n(z)$ is a product of n such factors and a constant unitary operator (the constant unitary factor, wherever it appears, can always be moved to the right). Then

$$K_B(w, z) = K_{B_0}(w, z) + B_0(z)K_{B_n}(w, z)B_0(w)^*.$$

It follows from the inductive hypothesis that this is a nonnegative kernel, and therefore $B(z)$ belongs to $\mathbf{S}_0(\mathfrak{G})$. If g_0 is a unit vector in the range of P_0, then $B_0(\alpha_0)^*g_0 = 0$, so that the function

$$h(z) = K_B(\alpha_0, z)g_0 = K_{B_0}(\alpha_0, z)g_0 = \frac{g_0}{1 - \bar{\alpha}_0 z}$$

belongs to both $\mathfrak{H}(B)$ and $\mathfrak{H}(B_0)$ and has the same norm in each space, and therefore $\mathfrak{H}(B_0)$ is contained isometrically in $\mathfrak{H}(B)$. It follows from Theorem 4.1.1 that $\mathfrak{H}(B_0)$ and $B_0\mathfrak{H}(B_n)$ are complementary spaces in $\mathfrak{H}(B)$ in the sense of de Branges. Since the inclusion of $\mathfrak{H}(B_0)$ in $\mathfrak{H}(B)$ is isometric, $B_0\mathfrak{H}(B_n)$ is also contained isometrically in $\mathfrak{H}(B)$,

(A4) $$\mathfrak{H}(B) = \mathfrak{H}(B_0) \oplus B_0\mathfrak{H}(B_n),$$

and multiplication by $B_0(z)$ maps $\mathfrak{H}(B_n)$ isometrically onto $B_0\mathfrak{H}(B_n)$. Now (1) is clear. We use Theorem 3.2.3(2) to establish (2). Assume that $B(z)g$ belongs to $\mathfrak{H}(B)$ for some $g \in \mathfrak{G}$. By (A4),

$$B(z)g = \frac{cg_0}{1 - \bar{\alpha}_0 z} + B_0(z)h_n(z).$$

for some scalar c and function $h_n(z)$ in $\mathfrak{H}(B_n)$. Since $B(z) = B_0(z)B_n(z)$, on multiplying on the left by P_0 and taking $z = \alpha_0$, we get $c = 0$. Thus $h_n(z) = B_n(z)g$ belongs to $\mathfrak{H}(B_n)$. By our inductive hypothesis and Theorem 3.2.3(2) applied to $\mathfrak{H}(B_n)$, $g = 0$. This proves (2), and the direct statement follows.

In the converse direction, the case $n = 1$ follows from Theorem A1 applied to $\mathfrak{P} = \mathfrak{H}(B)$ and Theorem 3.1.3 (which is applicable because $B(z)f \equiv 0$ only for $f = 0$ by (2) and Theorem 3.2.3(2)). Assume that the assertion is verified up to the dimension n, and suppose that $B(z)$ satisfies (1) and (2) with n replaced by $n+1$. Let $\bar{\alpha}_0$ be an eigenvalue for the transformation T which takes any $h(z)$ in $\mathfrak{H}(B)$ to $[h(z) - h(0)]/z$. An eigenfunction must have the form

$$h_0(z) = \frac{g_0}{1 - \bar{\alpha}_0 z}, \qquad 0 \neq g_0 \in \mathfrak{G}.$$

Without loss of generality, we may assume that $\|g_0\|_{\mathfrak{G}} = 1$. By Theorem 3.2.5, the difference-quotient identity

$$\left\langle \frac{h(z) - h(0)}{z}, \frac{h(z) - h(0)}{z} \right\rangle_{\mathfrak{H}(B)} = \langle h(z), h(z) \rangle_{\mathfrak{H}(B)} - \langle h(0), h(0) \rangle_{\mathfrak{G}}$$

holds in $\mathfrak{H}(B)$. Choosing $h(z) = h_0(z)$, we find that $|\alpha_0| < 1$ and

$$\|h_0\|_{\mathfrak{H}(B)}^2 = \frac{1}{1 - |\alpha_0|^2}.$$

Define $B_0(z)$ by (A3) with $\rho_0 = 1$ and $P_0 \in \mathfrak{L}(\mathfrak{G})$ the projection onto the span of g_0. Then $\mathfrak{H}(B_0)$ is contained isometrically in $\mathfrak{H}(B)$. Hence by Corollary 4.1.4, there is a factorization $B(z) = B_0(z)B_n(z)$ with $B_n(z)$ in $\mathbf{S}_0(\mathfrak{G})$ satisfying the equivalent conditions of Theorem 4.1.1. Since the inclusion of $\mathfrak{H}(B_0)$ in $\mathfrak{H}(B)$ is isometric, we have an orthogonal direct sum decomposition (A4), and multiplication by $B_0(z)$ is an isometry from $\mathfrak{H}(B_n)$ onto $B_0\mathfrak{H}(B_n)$. Therefore $B_n(z)$ satisfies conditions (1) and (2) of the theorem, and hence the converse statement follows by induction.

The isometric inclusion of such a space $\mathfrak{H}(B)$ in the Hardy class $H_{\mathfrak{G}}^2$ likewise follows by an inductive argument, beginning with Example 1 for the case of a simple Blaschke-Potapov factor and using (A4). $\qquad\square$

A parallel result holds for inverses of Blaschke products.

THEOREM A3. *Let $\mathfrak{G}$ be a Hilbert space, and let $S(z) = B(z)^{-1}$, where $B(z)$ is a Blaschke product of degree n with values in $\mathfrak{L}(\mathfrak{G})$ such that $B(0)$ is invertible. Then*

(1) *$S(z)$ belongs to $\mathbf{S}_n(\mathfrak{G})$ and $\mathfrak{H}(S)$ is the antispace of a Hilbert space of dimension n, and*

(2) *the canonical coisometric colligation V associated with $S(z)$ is unitary.*

Conversely, every function $S(z)$ satisfying (1) and (2) has this form.

Proof. Since $S(z) = B(z)^{-1}$,

$$K_S(w, z) = -S(z)K_B(w, z)S(w)^*.$$

By Theorem 1.5.7, multiplication by $S(z)$ is an isometry from $\mathfrak{H}(B)$ onto the antispace of $\mathfrak{H}(S)$, and so (1) holds by Theorem A2. To prove (2), we use Theorem 3.2.3(2). Suppose that $S(z)g$ belongs to $\mathfrak{H}(S)$ for some nonzero vector $g \in \mathfrak{G}$. Then the constant function $h_0(z) \equiv g$ belongs to $\mathfrak{H}(B)$ (some function in $\mathfrak{H}(B)$ is mapped to $S(z)g$ by multiplication by $S(z)$, and this is the only possibility). Thus the difference-quotient transformation in $\mathfrak{H}(B)$ has a nontrivial kernel, and so 0 is a null point of $B(z)$ according to the remarks following Theorem A2. This contradicts our assumption that $B(0)$ is invertible. Therefore $S(z)g$ belongs to $\mathfrak{H}(S)$ only for the vector $g = 0$, and hence (2) holds by Theorem 3.2.3(2).

Conversely, assume that $B(z)$ satisfies (1) and (2). When $n = 1$, the conclusion follows from Theorem A1. Assume that the converse statement holds up to the dimension n. Suppose that $S(z)$ satisfies (1) and (2) with n replaced by $n+1$. Let β_0 be an eigenvalue for the transformation T which takes any $h(z)$ in $\mathfrak{H}(S)$ to $[h(z) - h(0)]/z$. Then we can always choose an eigenfunction of the form

$$h_0(z) = \frac{g_0}{1 - \beta_0 z}, \qquad \|g_0\|_{\mathfrak{G}} = 1.$$

By Theorem 3.2.5, the difference-quotient identity holds in $\mathfrak{H}(S)$, and this gives

$$-\langle h_0, h_0 \rangle_{\mathfrak{H}(S)} = \frac{1}{|\beta_0|^2 - 1}.$$

In particular, $|\beta_0| > 1$. Write $\alpha_0 = 1/\beta_0$, and define

$$S_0(z) = B_0(z)^{-1}$$

where $B_0(z)$ is given by (A3) with this choice of α_0, $\rho_0 = 1$, and $P_0 \in \mathfrak{L}(\mathfrak{G})$ the projection onto the span of g_0. By Example 1(2), $\mathfrak{H}(S_0)$ is contained isometrically in $\mathfrak{H}(S)$. By Corollary 4.1.4,

$$S(z) = S_0(z)S_n(z)$$

where $S_n(z)$ belongs to $\mathbf{S}_n(\mathfrak{G})$. Multiplication by $S_0(z)$ maps $\mathfrak{H}(S_n)$ isometrically onto the orthogonal complement $S_0\mathfrak{H}(S_n)$ of $\mathfrak{H}(S_0)$ in $\mathfrak{H}(S)$. We easily check that $\mathfrak{H}(S_n)$ satisfies the conditions (1) and (2), and therefore the assertion follows by induction. $\qquad\square$

Suppose $S(z) = B(z)^{-1}$ as in Theorem A3, and let T be the main transformation in the canonical coisometric colligation, that is, $T : h(z) \to [h(z) - h(0)]/z$ in $\mathfrak{H}(S)$. Then the spectrum of T is given by

$$\sigma(T) = \{1/\alpha_1, \ldots, 1/\alpha_k\},$$

where $\alpha_1, \ldots, \alpha_k$ are the null points of $B(z)$. The hypothesis that $B(0)$ is invertible insures that $\alpha_j \neq 0$ for each $j = 1, \ldots, k$. By the proof of the theorem, we can justify this assertion in the following way. It is shown in the proof that any point in $\sigma(T)$ is the reciprocal of a null point of $B(z)$. Conversely, if α is a null point of $B(z)$, there is a nonzero vector $g \in \mathfrak{G}$ such that $B(\alpha)g = 0$. Then

$$\frac{B(z)}{z - \alpha} g = \frac{B(z) - B(\alpha)}{z - \alpha} g$$

belongs to $\mathfrak{H}(B)$ by §3.2. Since multiplication by $S(z) = B(z)^{-1}$ maps $\mathfrak{H}(B)$ onto $\mathfrak{H}(S)$ (see the proof of the theorem), the function $g/(z - \alpha)$ belongs to $\mathfrak{H}(S)$. This is an eigenfunction for T for the eigenvalue $1/\alpha$, and so $1/\alpha \in \sigma(T)$.

Some additional properties of finite-dimensional spaces used in Chapter 4 are noted in the next result.

THEOREM A4. *Let $\mathfrak{G}$ be a Hilbert space, and let $S(z) = B(z)^{-1}$, where $B(z)$ is a Blaschke product of finite degree with values in $\mathfrak{L}(\mathfrak{G})$ such that $B(0)$ is invertible. Then $\mathfrak{H}(S)$ has an algebraic basis consisting of a number of chains of the form*

$$\text{(A5)} \quad \begin{cases} u_0(z) = \dfrac{v_0}{1 - z/\alpha}, \\[2mm] u_1(z) = \dfrac{v_0 z}{(1 - z/\alpha)^2} + \dfrac{v_1}{1 - z/\alpha}, \\[2mm] \qquad \cdots \\[2mm] u_k(z) = \dfrac{v_0 z^k}{(1 - z/\alpha)^{k+1}} + \dfrac{v_1 z^{k-1}}{(1 - z/\alpha)^k} + \cdots + \dfrac{v_k}{1 - z/\alpha}, \end{cases}$$

where α is a null point of $B(z)$, $v_0, \ldots, v_k$ are vectors in $\mathfrak{G}$ with $v_0 \neq 0$, and $B(\alpha)v_0 = 0$. Every null point of $B(z)$ arises in some such chain.

The same point α may occur in more than one chain. In this case, the vectors v_0 that appear for different chains must then be linearly independent because the collection of all chains is an algebraic basis.

Proof. The spectrum of the transformation

$$T : h(z) \to [h(z) - h(0)]/z$$

in $\mathfrak{H}(S)$ is the set of numbers $1/\alpha$ where α is a null point of $B(z)$ (see the remarks following Theorem A3). Thus $\mathfrak{H}(S)$ has an algebraic basis relative to which the matrix of T is in Jordan canonical form. Such a basis consists of chains: to each eigenvalue $\beta = 1/\alpha$ of T there exist one or more chains $u_0, \ldots, u_k$ such that

$$Tu_0 = \beta u_0,$$
$$Tu_1 = \beta u_1 + u_0,$$
$$\cdots$$
$$Tu_k = \beta u_k + u_{k-1}.$$

The formulas (A5) exhibit a typical such chain. Since

$$K_B(w, z) = -B(z)K_S(w, z)B(w)^*,$$

multiplication by $B(z)$ is an isometry from the antispace of $\mathfrak{H}(S)$ onto $\mathfrak{H}(B)$. In particular,

$$B(z)u_0(z) = \frac{B(z)v_0}{1 - z/\alpha}$$

is analytic at $z = \alpha$, yielding $B(\alpha)v_0 = 0$. See Dym [1989a], pp. 49, 58, and Alpay and Dym [1986], Theorem 5.2, for further details. $\square$

We also note a uniqueness result.

THEOREM A5. *Let $\mathfrak{G}$ be a Hilbert space. Suppose that*

$$S_1(z) = B_1(z)^{-1} \qquad and \qquad S_2(z) = B_2(z)^{-1},$$

where $B_1(z)$ and $B_2(z)$ are Blaschke products of finite degree with values in $\mathfrak{L}(\mathfrak{G})$ such that $B_1(0)$ and $B_2(0)$ are invertible. If $\mathfrak{H}(S_1)$ and $\mathfrak{H}(S_2)$ coincide as sets, they are equal isometrically and

$$S_2(z) = S_1(z)U$$

for a constant unitary $U \in \mathfrak{L}(\mathfrak{G})$.

Proof. By assumption, $\mathfrak{H}(S_1)$ and $\mathfrak{H}(S_2)$ coincide as vector spaces. Denote this common vector space by $\mathfrak{P}$, and let Ω be the complex plane with the poles of the functions in $\mathfrak{P}$ deleted. The degrees of $B_1(z)$ and $B_2(z)$ are equal and coincide with $n = \dim \mathfrak{P}$ by Theorem A2.

Fix an algebraic basis $h_1, \ldots, h_n$ for $\mathfrak{P}$. If P is a nonpositive and invertible $n \times n$ matrix, then

$$\left\langle \sum_{j=1}^n \xi_j h_j, \sum_{i=1}^n \eta_i h_i \right\rangle_P = \langle P\xi, \eta \rangle_{\mathbf{C}^n}, \qquad \xi = \begin{pmatrix} \xi_1 \\ \vdots \\ \xi_n \end{pmatrix}, \ \eta = \begin{pmatrix} \eta_1 \\ \vdots \\ \eta_n \end{pmatrix}$$

defines an inner product on $\mathfrak{P}$ which makes the space into the antispace of a Hilbert space. Conversely, if $\langle \cdot, \cdot \rangle$ is an inner product with this property, then $\langle \cdot, \cdot \rangle = \langle \cdot, \cdot \rangle_P$ for a unique nonnegative and invertible $n \times n$ matrix P; in fact,

$$P = \left(\langle h_j, h_i \rangle \right)_{i,j=1}^n$$

is the Gram matrix of $h_1, \ldots, h_n$. Let T be the difference-quotient transformation in $\mathfrak{P}$. We shall prove the theorem by showing that there is exactly one nonpositive and invertible matrix P such that

(A6) $\qquad \langle Th, Th \rangle_P = \langle h, h \rangle_P - \langle h(0), h(0) \rangle_{\mathfrak{G}}, \qquad h \in \mathfrak{P}.$

This identity also holds when P is chosen so that $\langle \cdot, \cdot \rangle_P$ is the inner product of $\mathfrak{H}(S_1)$ or $\mathfrak{H}(S_2)$ by Theorems A3 and 3.2.5. Once we have verified the assertion, it will follow that $\mathfrak{H}(S_1)$ and $\mathfrak{H}(S_2)$ are equal isometrically. Equality of the reproducing kernels for these spaces implies that

$$B_1(z)^{-1} B_1(w)^{*-1} = B_2(z)^{-1} B_2(w)^{*-1}, \qquad w, z \in \Omega.$$

Since $B(z)$ has unitary values on the unit circle,

$$B_1(z) = U B_2(z)$$

for a constant unitary operator $U \in \mathfrak{L}(\mathfrak{G})$, yielding the result.

For each $w \in \Omega$, define an operator $E(w) \in \mathfrak{L}(\mathbf{C}^n, \mathfrak{G})$ by

$$E(w)\xi = h_1(w)\xi_1 + \cdots + h_n(w)\xi_n, \qquad \xi = \begin{pmatrix} \xi_1 \\ \vdots \\ \xi_n \end{pmatrix} \in \mathbf{C}^n.$$

For each $\xi \in \mathbf{C}^n$, $E(z)\xi$ belongs to $\mathfrak{P}$ as a function of z, and

(A7) $\qquad \langle E(\cdot)\xi, E(\cdot)\eta \rangle_P = \langle P\xi, \eta \rangle_{\mathbf{C}^n}, \qquad \xi, \eta \in \mathbf{C}^n.$

Since $\mathfrak{P}$ is invariant under T, there is an $n \times n$ matrix A such that

(A8) $\qquad \dfrac{E(z)\xi - E(0)\xi}{z} = E(z)A\xi, \qquad \xi \in \mathbf{C}^n.$

As a linear transformation on $\mathbf{C}^n$, A is similar to T. The spectrum of T and hence the spectrum of A lies in the complement of the closure of the unit disk $\mathbf{D}$. Writing $C = E(0)$, we obtain

$$E(z) = C(1 - zA)^{-1}.$$

In (A6), choose $h(z) = E(z)\xi$ for some $\xi \in \mathbf{C}^n$. Then by (A7) and (A8),

$$\langle PA\xi, A\xi \rangle_{\mathbf{C}^n} = \langle P\xi, \xi \rangle_{\mathbf{C}^n} - \langle C\xi, C\xi \rangle_{\mathfrak{G}},$$

and so $A^*PA - P = -C^*C$. Equivalently,

$$P - A^{*-1}PA^{-1} = -A^{*-1}C^*CA^{-1}.$$

Since the spectrum of A^{-1} lies in $\mathbf{D}$, this equation has a unique solution by a standard iterative method: $P = -\sum_{k=1}^{\infty} A^{*-k}C^*CA^{-k}$. The series converges because

$$\left\| A^{-k} \right\|^{1/k} < \rho$$

for all sufficiently large k and some $\rho < 1$. Hence (A6) holds for exactly one inner product on $\mathfrak{P}$ which makes the space into the antispace of a Hilbert space. $\quad\square$

NOTES

The notes and bibliography do not aim to be complete. However, a work such as the present one obviously owes a great debt to previous authors. We credit some of these sources and point to additional literature.

Chapter 1

§1.1. Accounts of operator theory on Kreĭn spaces may be found in Azizov and Iokhvidov [1986], Bognár [1974], Dritschel and Rovnyak [1990,1996], and Iokhvidov, Kreĭn, and Langer [1982].

A classical account of reproducing kernel Hilbert spaces is Aronszajn [1950]. The first general treatments of the nonpositive case are Schwartz [1964] and Sorjonen [1975]. Theorem 1.1.4 seems to be new. The idea of the proof of Lemma 1.1.5 is in Alpay and Dym [1986], Theorem 2.1; an analogous result for the scalar case is given in Donoghue [1974], Lemma 3 on p. 143.

§1.2. A good survey of the theory of colligations is Brodskiĭ [1978]. Kreĭn space generalizations are given in Ćurgus, Dijksma, Langer, and de Snoo [1989] and Dijksma, Langer, and de Snoo [1986a]. Characteristic functions are also called transfer functions; for their use in linear system theory, see, for example, Arov [1979], Bart, Gohberg, and Kaashoek [1979], Fuhrmann [1981], Helton [1974], Kaashoek [1996], and Kailath [1980]. The concepts of closely inner and outer connected colligations occur in mathematical system theory, where the terms "controllable" and "observable" are used. We briefly touch upon these notions in §3.5 A.

§1.3. Theorem 1.3.1 was called to our attention by A. Gheondea. Theorem 1.3.2 is similar to results in Brodskiĭ [1978], Ćurgus, Dijksma, Langer, and de Snoo [1989] (Lemma 3.1), and Dijksma, Langer, and de Snoo [1986a] (Propositions 3.1, 3.2). The Bognár-Krámli factorization (1.3.1) is given in Bognár [1974], Theorem 2.1 on p. 149, and Dritschel and Rovnyak [1996], Theorem 1.1. The existence of Julia operators was first proved by Arsene, Constantinescu, and Gheondea [1987], which is also the source for the method to prove Theorem 1.3.4 using (1.3.14) and (1.3.15). Dritschel [1993] gives a complete analysis of the essential uniqueness question. Concerning defect and Julia operators generally, see Constantinescu and Gheondea [1993], Ch. Davis [1970], and Dritschel and Rovnyak [1990,1996].

§1.4. The proof of Theorem 1.4.1 is due to T. Ya. Azizov and was communicated privately. Two proofs of the more general extension theorem of Shmul'yan for Kreĭn space operators are given in Dritschel and Rovnyak [1990], Theorem 1.4.4, and Dritschel and Rovnyak [1996], the supplement. The present version for relations in Theorem 1.4.2 is new; another proof is given in Alpay, Dijksma, Rovnyak, and de Snoo [1997]. The relations version also has a generalization to Kreĭn spaces, given in Dritschel and Rovnyak [1996], the supplement. For the calculus of linear relations, see Arens [1961]. Weak isomorphisms and more general conditions which imply the existence of a continuous extension are discussed in Ćurgus, Dijksma, Langer, and de Snoo [1989].

§1.5 A. The theory of contractively contained Kreĭn spaces and their complementation properties is due to de Branges [1988a]; an account by the author of the theory is also given in de Branges [to appear]. The present approach follows Dritschel and Rovnyak [1990,1991]. Related operator methods may also be found in Constantinescu and Gheondea [1993], Ćurgus and Langer [1990], and Hara [1992].

§1.5 B. Theorems 1.5.5 and 1.5.6 are apparently new. The proof of (1.5.5) is due to S. O. Hassi; the inequality also follows from the minimax principle or Lemma 1.1.1' (whose proof uses the minimax principle).

Chapter 2

§2.1. Kernels of the form $K_S(w, z), K_{\tilde{S}}(w, z), D_S(w, z)$ and their close relatives are well known in operator theory and its applications. They are sometimes called Nevanlinna-Pick or Schwarz-Pick kernels. In a series of papers, Kreĭn and Langer [1971ab,1972,1973,1977–1985,1981] studied the structural properties of operator-valued functions for which such kernels have a finite number of negative squares. Kernels having a finite number of negative squares also arise in the Takagi problem and are treated in Adamjan, Arov, and Kreĭn [1971].

Systematic use of the spaces $\mathfrak{H}(S), \mathfrak{H}(\tilde{S}), \mathfrak{D}(S)$ is a feature of the canonical model of de Branges and Rovnyak [1966a,1966b]. The coordinate-free approach of Nikol'skiĭ and Vasyunin [1989] accounts for this model and the model of Sz.-Nagy and Foiaş [1970]. The viewpoint of colligations and linear systems evolved in the 1970's and 1980's and appears in Ball and Cohen [1991] and de Branges [1985]. Theorems 2.1.2 and 2.1.3 extend results in Ball [1975,1978], Brodskiĭ [1978], and McEnnis [1979]. For another approach to characteristic functions and canonical models, see Kuzhel [1996]. Generalizations to operator vessels and several nonselfadjoint operators can be found in Livšic, Kravitsky, Markus, and Vinnikov [1995].

An equivalent form of the kernel $K_S(w, z)$ is $[J - S(z)JS(w)^*]/(1 - z\bar{w})$, where J is a signature operator on a Hilbert space. The study of such kernels in the nonnegative case ($\kappa = 0$) goes back to Potapov [1955]. A survey with references and applications may be found in the CBMS lectures of Dym [1989a]. The "ϵ-method" of Sz.-Nagy and Korányi [1958] for nonnegative kernels is generalized to kernels with negative squares in Kreĭn and Langer [1971ab,1972]. The study of kernels $K_S(w, z)$ is frequently associated with interpolation, as in Ball [1983], Ball, Gohberg, and Rodman [1990], and Ball and Helton [1982,1983]. The approach of the Potapov school can be seen in Golinskiĭ [1983] and Kheifets and Yuditskiĭ [1994]. Another method is given in Dijksma and Langer [to appear]. In the scalar case, Pontryagin spaces $\mathfrak{H}(S)$ are used in expansions of functions of several variables by Clark [1990].

§2.2–2.3. Theorems 2.2.1, 2.2.2, and 2.3.1 extend results of de Branges and Rovnyak [1966a,1966b] in the Hilbert space case. This approach, which is based on the extension theorem for relations in Theorem 1.4.2, saves separate verification of special properties, such as invariance under the difference-quotient transformation. Instead, such properties are consequences of the construction. Filimonov [1977] solves a related realization problem involving kernels $K_S(w, z)$ having a finite number of negative squares and Pontryagin state spaces.

Kreĭn space extensions of the main realization theorems have previously been given in de Branges [1988b,1991,1994,to appear] and Yang [1994] by methods based on complementation theory. The realization problem is also treated in the indefinite setting in Azizov [1982,1984], Dijksma, Langer, and de Snoo [1986a,1986b,1987a], and Marcantognini [1990]. Azizov [1982] gave necessary and sufficient conditions for a function $S(z)$ in $\mathbf{H}_0(\mathfrak{F}, \mathfrak{G})$ to be the characteristic function of a unitary colligation whose state space is a Pontryagin space with specified negative index in terms of the coefficients in the Taylor expansion $S(z) = \sum_{n=0}^{\infty} S_n z^n$; see Azizov and Iokhvidov [1986], Theorem 3.16 on p. 275 and the notes on p. 285.

Coisometric, isometric, and unitary realizations are constructed in Alpay, Bolotnikov, Dijksma, and de Snoo [1993,1996]. The kernels here are similar, except that expressions of the form $a(z)\overline{a(w)} - b(z)\overline{b(w)}$ are used in place of denominators $1 - z\bar{w}$. For an adaptation to a nonstationary setting, see Alpay and Peretz [1996].

§2.4. Theorems 2.4.1 and 2.4.3 extend results in Dijksma, Langer, and de Snoo [1986a] for the class $\mathbf{S}_\kappa(\mathfrak{F}, \mathfrak{G})$ when $\mathfrak{F}$ and $\mathfrak{G}$ are Hilbert spaces.

§2.5 A. The classical source for the classes $\mathbf{S}_\kappa(\mathfrak{F}, \mathfrak{G})$ is Schur [1917,1918]. The first result on the relationship of the number of negative squares of the kernels $K_S(w, z), K_{\tilde{S}}(w, z), D_S(w, z)$ is that of de Branges and Rovnyak [1966b]: when $\mathfrak{F}, \mathfrak{G}$ are Hilbert spaces, the three kernels are simultaneously nonnegative or not. Generalizations to nonnegative kernels with Kreĭn space coefficient spaces are due to Andô [1990], Ball [1975], and McEnnis [1979]. When $\mathfrak{F}$ and $\mathfrak{G}$ are Hilbert spaces, Dijksma, Langer, and de Snoo [1986a] proved that the three kernels have the same number of negative squares. In Alpay, Dijksma, van der Ploeg, and de Snoo [1992], this result is extended to the case where $\mathfrak{F}$ and $\mathfrak{G}$ are Pontryagin spaces having the same index, which is our Theorem 2.5.2. The remarks preceding Theorem 2.5.3 on the values of generalized Schur functions are related to a maximum principle discussed in Dijksma, Langer, and de Snoo [1986a], Proposition 8.1. The necessity part of Theorem 2.5.5 also follows from interpolation methods, as in Rosenblum and Rovnyak [1982].

§2.5 B. The realization theorems for base points α other than the origin are based on Alpay, Bolotnikov, Dijksma, and de Snoo [1993,1996]. The forms of reproducing kernels used there are more general. The method based on Theorem 2.5.7 is new.

§2.5 C. The method used here is applied by Ball and Trent [1996] to the Toeplitz corona theorem and generalizations.

§2.5 D. This example was called to our attention by V. E. Katsnelson and N. Young.

Chapter 3

§3.1. Theorem 3.1.2 on the characterization of spaces $\mathfrak{H}(S)$ is related to a theorem of Guyker [1995] and to the study of shift-invariant spaces and Beurling-Lax theorems. For example, see Halmos [1961], Helson [1964], Rosenblum and Rovnyak [1985], and Sand [1995] for the Hilbert space case, and Ball and Helton [1983], de Branges [to appear], and Möller [1991] for Kreĭn space generalizations. In the scalar case, characterizations of spaces $\mathfrak{H}(S)$ are given in Guyker [1991] and Leech [1969]. A characterization of spaces $\mathfrak{D}(S)$ can be found in de Branges [1970], Theorem 4, for functions in $\mathbf{S}_0(\mathfrak{F})$ with $\mathfrak{F}$ a Hilbert space.

§3.2–3.3. These results generalize theorems for the canonical model in the scalar and Hilbert space cases.

A key issue is if equality always holds in (3.1.1), and Theorems 3.2.4–3.2.5 and 3.3.4–3.3.5 give special properties of the state spaces when this occurs. In the scalar and Hilbert space situations, the case of equality was characterized by Ball and Kriete [1987], Nikol'skiĭ and Vasyunin [1986], and Sarason [1986a]. The scalar theory has been developed in directions not discussed in our account in other places as well, such as Davis and McCarthy [1991] and Lotto and Sarason [1993], for example; see Sarason [1994] for a recent survey.

The Ginzburg inequality (3.2.12) is given in Kreĭn and Langer [1972] and is used by Brodskiĭ [1978] to construct a unitary realization of a function in $\mathbf{S}_0(\mathfrak{F},\mathfrak{G})$ when $\mathfrak{F},\mathfrak{G}$ are Hilbert spaces. In Dijksma, Langer, and de Snoo [1986a], the inequality is derived as a consequence of such a realization. The present version is new for the setting of Pontryagin spaces having the same negative index.

§3.4. Theorem 3.4.3 is an example of a result of de Branges [1988a] which constructs a Kreĭn space associated with any selfadjoint operator P such that $P^2 \leq tP$ for some $t > 0$. Similar model spaces are constructed as operator ranges in Nikol′skiĭ and Vasyunin [1989] and Pták and Vrbová [1993]. The interpretation of $\mathfrak{D}(S)$ as a space of holomorphic functions on $\Omega_+ \cup \Omega_-$ is similar to an argument in Kreĭn and Langer [1972]; see Dijksma, Langer, and de Snoo, [1986a], p. 149.

§3.5 A. The present methods give another approach to the results of Alpay and Gohberg [1988]. Finite-dimensional spaces are also studied in Alpay and Dym [1986]. The literature on this subject is large. For example, see also Genin, Van Dooren, Kailath, Delosme, and Morf [1983] and Glover [1984].

§3.5 B. The condition $S(z) = V^*S(-z)U$ occurs in the study of periodic systems; for example, see Alpay, Bolotnikov, and Loubaton [to appear]. The results of this subsection can be extended to different symmetries including a rotation through an angle or a conformal mapping of the unit disk onto itself.

§3.5 C. For another proof of Leech's theorem, see Rosenblum and Rovnyak [1985], Examples and Addenda at the end of Chapter 5; there, however, it is assumed that $A(z)$ and $B(z)$ are holomorphic and bounded on the unit disk. Theorem 3.5.7 generalizes a scalar result of Sarason [1990]. Similar applications of Leech's theorem are given in Alpay, Bolotnikov, and Peretz [1995]. A connection between Leech's theorem and a theorem of Shmul′yan [1967] is shown in Dritschel and Rovnyak [1990], §3.4. Theorem 3.5.9 is related to Sarason [1986b], Theorem 2; the condition $S\mathfrak{H}(S) \subseteq \mathfrak{H}(S)$ occurs there in the scalar case, but with the extra condition that $S(z)$ is in the nonextreme point case. Our result characterizes the inclusion $S\mathfrak{H}(S) \subseteq \mathfrak{H}(S)$ with no consideration of case.

§3.5 D. Theorem 3.5.10 extends Problem 90 of de Branges and Rovnyak [1966a]. For other generalizations and related results, see Ball and Kriete [1987], Nikol′skiĭ and Vasyunin [1986], and Sarason [1986a,1994].

Chapter 4

§4.1 A. The relationship between factorization and invariant subspaces originates in Livšic and Potapov [1950] and is one of the most fundamental aspects of the theory of characteristic operator functions and canonical models. The property $\kappa = \kappa_1 + \kappa_2$ in factorizations of the type given in Theorem 4.1.1 was noted by Sakhnovich [1986] (see Chapter II, Theorem 2.3, p. 25 of the English translation). The theorems of this section for the spaces $\mathfrak{H}(S)$ may be viewed as generalizing results in de Branges and Rovnyak [1966a,1966b] to the Schur classes $\mathbf{S}_\kappa(\mathfrak{F}, \mathfrak{G})$.

Example 1 is studied in greater depth by Dijksma, Langer, and de Snoo [1986a], §10, when the coefficient spaces are Hilbert spaces. The example can be iterated, leading to the notion of an extension space as used in Alpay [1989] in a Kreĭn space version of the commutant lifting theorem. When $\mathfrak{F}$ and $\mathfrak{G}$ are Hilbert spaces, Schur functions $S(z)$ with $S(0) = 0$ have a unitary colligation with the main operator a partial isometry. This corresponds to the theory of characteristic functions of symmetric operators as studied by Štraus [1960,1968].

§4.1 B. The factorization theory for the spaces $\mathfrak{D}(S)$ is due to de Branges [1970,1994].

§4.2 A. The original source for the Kreĭn-Langer factorization is Kreĭn and Langer [1972], Theorem 3.2 on p. 382. Another proof may be found in Dijksma, Langer, and de Snoo [1986a].

§4.2 B. Strongly regular factorizations are discussed in Dijksma, Langer, and de Snoo [1986a], §7, and Kreĭn and Langer [1972,1981]. Regular factorizations in the sense of Brodskiĭ [1978] are obtained from an arbitrary triangularization of the difference-quotient transformation. This notion differs from our definition, and hence we use the term "strongly regular" to distinguish it. Strong regularity implies regularity in Brodskiĭ's sense. It corresponds to a special triangular form of T. For further discussion, see the remarks after the proof of Proposition 7.10 in Dijksma, Langer, and de Snoo [1986a]. The argument for Lemma 4.2.8 is adapted from de Branges and Rovnyak [1966b], Appendix (this occurs on p. 387 in the proof of Theorem 17).

§4.3. The Potapov-Ginzburg transform is also called the Redheffer transform and is related to Schur complements; see Arov [1973], Dym [1989a], and Iokhvidov, Kreĭn, and Langer [1982]. The proof of Theorem 4.3.3(1) is adapted from Alpay, Dijksma, van der Ploeg, and de Snoo [1992], Theorem 3.4.

The results of §4.2 and §4.3 give partial information on the structure of functions in the classes $\mathbf{S}_\kappa(\mathfrak{F}, \mathfrak{G})$. In special cases, much more can be said. Potapov [1955] gives the multiplicative structure of matrix-valued functions in $\mathbf{S}_0(\mathfrak{G})$ when $\mathfrak{G}$ is a finite-dimensional Kreĭn space. Ginzburg [1967a,b] has generalized these results to arbitrary Kreĭn spaces with additional hypotheses involving compactness or invertibility of function values. When $\mathfrak{F}$ and $\mathfrak{G}$ are Hilbert spaces, the boundary behavior of functions in $\mathbf{S}_\kappa(\mathfrak{F}, \mathfrak{G})$ can be studied by means of colligations. For $\kappa = 0$, this was done by Štraus [1965,1966] and for $\kappa > 0$ by Dijksma, Langer, and de Snoo [1986a]. The class of functions which are of bounded type on $\mathbf{D}$ is sometimes called the Nevanlinna class for $\mathbf{D}$; see, for example, Rosenblum and Rovnyak [1994]. Function theory for matrix-valued holomorphic functions is surveyed in Katsnelson and Kirstein [to appear].

§4.4 A,B,C. Theorem 4.1.1(1), Lemma 4.4.2, and Lemma 4.4.3 are due to Alpay, Dijksma, van der Ploeg, and de Snoo [1992]. In that paper it is proved for $S(z)$ in $\mathbf{H}_0(\mathfrak{F}, \mathfrak{G})$ with $\mathfrak{F}, \mathfrak{G}$ Kreĭn spaces, that $\text{sq}_- \mathfrak{D}_S$ is finite if and only if $\text{sq}_- K_S$ and $\text{sq}_- K_{\tilde{S}}$ are finite; there are also refinements when $\mathfrak{F}$ and $\mathfrak{G}$ are Pontryagin spaces. The other results in this section are new for the present generality.

§4.5. The theorems on models imply, in particular, that the given operator is determined up to unitary equivalence by its characteristic function. Štraus [1960] defined the characteristic function of a bounded operator which is not necessarily a contraction and showed that its completely nonunitary part is determined up to unitary equivalence by its characteristic function. In the indefinite case, operators satisfying the condition $\text{ind}_- (1 - T^*T) < \infty$ are studied in Gheondea [1993]. Functions of the type in Example 2 are sometimes called Potapov factors of the third kind. They occur in iterative procedures similar to the Schur algorithm in mathematical system theory, where the term "Brune section" is also used; see Delsarte and Genin [1992] and Dewilde and Dym [1984].

Epilogue

The recent book by Constantinescu [1996] gives an excellent overview of the Schur algorithm in its many forms in operator theory and applications. A proof of Theorem E2 may be found in Christner and Rovnyak [1995]. A related result on Taylor coefficients due to Azizov [1982] appears in Azizov and Iokhvidov [1986], Theorem 3.16 on p. 275. See Constantinescu and Gheondea [1996a,b,c] for recent progress on extension problems and associated Hermitian kernels related to the Schur algorithm. Generalizations of the Schur algorithm and linear fractional representations appear in Alpay and Dym [1986], Arov and Grossman [1983,1992], Dijksma, Marcantognini, and de Snoo [1994], and Dubovoj, Fritzsche, and Kirstein [1992]. An early work in this direction is Chamfy [1958]. For applications of the Schur algorithm and its generalizations to mathematical engineering, see the survey of Kailath and Sayed [1995].

Nevanlinna functions appear in de Branges [1968], Dym [1994b], and Kreĭn and Langer [1977–1985], Part I, for example. Their relation to Schur functions is studied in Bruinsma, Dijksma, and de Snoo [1993], Dijksma, Langer, and de Snoo [1987b], and Kreĭn and Langer [1981].

Appendix

The Liapunov-Stein equation in the proof of Theorem A5 is related to work of Alpay and Gohberg [1988] and Dym [1989b]. The equation is an equivalent form of a resolvent identity that characterizes the metric structure of reproducing kernel spaces of holomorphic functions. This equivalence seems to have been first pointed out by Dym [1989b]. In resolvent form, the identity is due to de Branges [1963–1965]; Rovnyak [1968] added a technical improvement and Ball [1975] put the result in a disk setting. See Alpay and Dym [1993] and Dym [1994a] for accounts; the latter contains applications to interpolation theory.

REFERENCES

V. M. Adamjan, D. Z. Arov, and M. G. Kreĭn [1971], *Analytic properties of the Schmidt pairs of a Hankel operator and the generalized Schur-Takagi problem*, Mat. Sb. (N.S.) **86**(128), 34–75; English transl.: Math. USSR-Sb. **15** (1971), no. 1, 31–73; MR 45#7505.

D. Alpay [1989], *Dilatations des commutants d'opérateurs pour des espaces de Kreĭn de fonctions analytiques*, Ann. Inst. Fourier (Grenoble) **39**, no. 4, 1073–1094; MR 91a:47045.

D. Alpay, V. Bolotnikov, A. Dijksma, and H. S. V. de Snoo [1993], *On some operator colligations and associated reproducing kernel Hilbert spaces*, Oper. Theory: Adv. Appl., vol. 61, Birkhäuser, Basel, pp. 1–27; MR 94i:47018.

―― [1996], *On some operator colligations and associated reproducing kernel Pontryagin spaces*, J. Funct. Anal. **136**, 39–80; MR 97a:47014.

D. Alpay, V. Bolotnikov, and Ph. Loubaton [to appear], *Dissipative periodic systems and symmetric interpolation in Schur classes*, Arch. Math. (Basel).

D. Alpay, V. Bolotnikov, and Y. Peretz [1995], *On the tangential interpolation problem for H_2 functions*, Trans. Amer. Math. Soc. **347**, 675–686; MR 95e:47015.

D. Alpay, A. Dijksma, J. van der Ploeg, and H. S. V. de Snoo [1992], *Holomorphic operators between Kreĭn spaces and the number of squares of associated kernels*, Oper. Theory: Adv. Appl., vol. 59, Birkhäuser, Basel, pp. 11–29; MR 95g:47056.

D. Alpay, A. Dijksma, J. Rovnyak, and H. S. V. de Snoo [1996], *Fonctions de Schur, colligations d'opérateurs, et espaces de Pontryagin à noyau reproduisant*, C. R. Acad. Sci. Paris Sér. I Math. **322**, no. 1, 15–20; MR 97d:47018.

―― [1997], *Reproducing kernel Pontryagin spaces*, Holomorphic Spaces, eds. S. Axler, J. McCarthy, and D. Sarason; MSRI Publications, vol. 33, Cambridge University Press, New York.

D. Alpay and H. Dym [1986], *On applications of reproducing kernel spaces to the Schur algorithm and rational J unitary factorization*, Oper. Theory: Adv. Appl., vol. 18, Birkhäuser, Basel, pp. 89–159; MR 89g:46051.

―― [1993], *On a new class of structured reproducing kernel spaces*, J. Funct. Anal. **111**, 1–28; MR 94g:46035.

D. Alpay and I. Gohberg [1988], *Unitary rational matrix functions*, Oper. Theory: Adv. Appl., vol. 33, Birkhäuser, Basel, pp. 175–222; MR 90m:47006.

D. Alpay and Y. Peretz [1996], *Special realizations for Schur upper triangular operators*; preprint.

T. Andô [1990], *de Branges Spaces and Analytic Operator Functions*, Hokkaido University, Research Institute of Applied Electricity, Division of Applied Mathematics, Sapporo.

R. Arens [1961], *Operational calculus of linear relations*, Pacific J. Math. **11**, 9–23; MR 23#A517.

N. Aronszajn [1950], *Theory of reproducing kernels*, Trans. Amer. Math. Soc. **68**, 337–404; MR 14,479c.

D. Z. Arov [1973], *Realization of matrix-valued functions according to Darlington*, Izv. Akad. Nauk SSSR Ser. Mat. **37**, 1299–1331; English transl.: Math. USSR-Izv. **7** (1973), 1295–1326; MR 50#10287.

_____ [1979], *Passive linear steady-state dynamical systems*, Sibirsk. Mat. Zh. **20**, no. 2, 211–228, 457; English transl.: Siberian Math. J. **20** (1979), no. 2, 149–162; MR 80g:93031.

D. Z. Arov and L. Z. Grossman [1983], *Scattering matrices in the theory of extensions of isometric operators*, Dokl. Akad. Nauk. SSSR **270**, no. 1, 17–20; English transl.: Soviet Math. Dokl. **27** (1983), 518–522; MR 85c:47008.

_____ [1992], *Scattering matrices in the theory of unitary extensions of isometric operators*, Math. Nachr. **157**, 105–123; MR 94i:47019.

Gr. Arsene, T. Constantinescu, and A. Gheondea [1987], *Lifting of operators and prescribed numbers of negative squares*, Michigan Math. J. **34**, 201–216; MR 88j:47008.

T. Ya. Azizov [1982], *On the theory of extensions of isometric and symmetric operators in spaces with an indefinite metric*, Technical Report 3420-82, Voronezh University.

_____ [1984], *On the theory of extensions of J-isometric and J-symmetric operators*, Funktsional. Anal. i Prilozhen. **18**, 57–58; English transl.: Funct. Anal. Appl. **18** (1984), 46–48; MR 85k:47061.

T. Ya. Azizov and I. S. Iokhvidov [1986], *Foundations of the Theory of Linear Operators in Spaces with an Indefinite Metric*, "Nauka", Moscow; English transl.: Wiley, New York, 1989; MR 88g:47070, 90j:47042.

J. A. Ball [1975], *Models for noncontractions*, J. Math. Anal. Appl. **52**, 235–254; MR 52#8972.

_____ [1978], *Factorization and Model Theory for Contraction Operators with Unitary Part.*, Mem. Amer. Math. Soc. **13**, no. 198; MR 80d:47013.

_____ [1983], *Interpolation problems of Pick-Nevanlinna and Loewner types for meromorphic matrix functions*, Integral Equations and Operator Theory **6**, 804–840; MR 85k:30054.

J. A. Ball and N. Cohen [1991], *de Branges-Rovnyak operator models and systems theory: a survey*, Oper. Theory: Adv. Appl., vol. 50, Birkhäuser, Basel, pp. 93–136; MR 93a:47011.

J. A. Ball, I. Gohberg, and L. Rodman [1990], *Interpolation of Rational Matrix Functions*, Oper. Theory: Adv. Appl., vol. 45, Birkhäuser, Basel; MR 92m:47027.

J. A. Ball and J. W. Helton [1982], *Lie groups over the field of rational functions, signed spectral factorization, signed interpolation, and amplifier design*, J. Operator Theory **8**, 19–64; MR 84j:94033.

____ [1983], *A Beurling-Lax theorem for the Lie group $U(m, n)$ which contains most interpolation theory*, J. Operator Theory **9**, 107–142; MR 84m:47046.

J. A. Ball and T. L. Kriete III [1987], *Operator-valued Nevanlinna-Pick kernels and the functional models for contraction operators*, Integral Equations and Operator Theory **10**, 17–61; MR 88a:47013.

J. A. Ball and T. Trent [1996], *Unitary colligations, reproducing kernel Hilbert spaces, and Nevanlinna-Pick interpolation in several variables*; preprint.

H. Bart, I. Gohberg, and M. A. Kaashoek [1979], *Minimal Factorization of Matrix and Operator Functions*, Oper. Theory: Adv. Appl., vol. 1, Birkhäuser, Basel; MR 81a:47001.

J. Bognár [1974], *Indefinite Inner Product Spaces*, Ergeb. Math. Grenzgeb., Bd. 78, Springer-Verlag, New York-Heidelberg; MR 57#7125.

L. de Branges [1963–1965], *Some Hilbert spaces of analytic functions. I*, Trans. Amer. Math. Soc. **106** (1963), 445–468; MR 26#2866; II, J. Math. Anal. Appl. **11** (1965), 44–72; III, ibid. **12** (1965), 149–186; MR 35#778,779.

____ [1968], *Hilbert Spaces of Entire Functions*, Prentice-Hall, Englewood Cliffs, NJ; MR 37#4590.

____ [1970], *Factorization and invariant subspaces*, J. Math. Anal. Appl. **29**, 163–200; MR 40#7846.

____ [1985], *Nodal Hilbert spaces of analytic functions*, J. Math. Anal. Appl. **108**, 447–465; MR 87i:47020.

____ [1988a], *Complementation in Kreĭn spaces*, Trans. Amer. Math. Soc. **305**, 277–291; MR 89c:46034.

____ [1988b], *Kreĭn spaces of analytic functions*, J. Funct. Anal. **81**, 219–259; MR 90c:47060.

____ [1991], *A construction of Kreĭn spaces of analytic functions*, J. Funct. Anal. **98**, 1–41; MR 92k:46031.

____ [1994], *Factorization in Kreĭn spaces*, J. Funct. Anal. **124**, 228–262; MR 95f:47032.

____ [to appear], *Square Summable Power Series*, Springer-Verlag, New York-Heidelberg.

L. de Branges and J. Rovnyak [1966a], *Square Summable Power Series*, Holt, Rinehart and Winston, New York; MR 35#5909.

—— [1966b], *Canonical models in quantum scattering theory*, Perturbation Theory and its Applications in Quantum Mechanics (Proc. Adv. Sem. Math. Res. Center, U.S. Army, Theoret. Chem. Inst., Univ. of Wisconsin, Madison, Wis., 1965), pp. 295–392; MR 39#6109.

M. S. Brodskiĭ [1978], *Unitary operator colligations and their characteristic functions*, Uspekhi Mat. Nauk **33**, no. 4(202), 141–168, 256; English transl.: Russian Math. Surveys **33** (1978), no. 4, 159–191; MR 80e:47010.

P. Bruinsma, A. Dijksma, and H. S. V. de Snoo [1993], *Models for generalized Carathéodory and Nevanlinna functions*, Challenges of a Generalized System Theory (Amsterdam, 1992), Konink. Nederl. Akad. Wetensch. Verh. Afd. Natuurk. Eerste Reeks, vol. 40, North-Holland, Amsterdam, pp. 161–178; MR 96e:47013.

C. Chamfy [1958], *Fonctions méromorphes dans le cercle-unité et leurs séries de Taylor*, Ann. Inst. Fourier (Grenoble) **8**, 211–262; MR 21#6433.

G. Christner [1993], *Applications of the Extension Properties of Operators on Kreĭn Spaces*, Dissertation, University of Virginia.

G. Christner and J. Rovnyak [1995], *Julia operators and the Schur algorithm*, Harmonic Analysis and Operator Theory (Caracas, 1994), Contemp. Math., vol. 189, Amer. Math. Soc., Providence, RI, pp. 135–160; MR 96h:47022.

D. N. Clark [1990], *The Kreĭn space $(\varphi H^2)^\perp$ and coefficients of analytic functions*, Oper. Theory: Adv. Appl., vol. 43, Birkhäuser, Basel, pp. 131–140; MR 93b:46047.

T. Constantinescu [1996], *Schur Parameters, Factorization and Dilation Problems*, Oper. Theory: Adv. Appl., vol. 82, Birkhäuser, Basel.

T. Constantinescu and A. Gheondea [1993], *Elementary rotations of linear operators in Kreĭn spaces*, J. Operator Theory **29**, 167–203; MR 95b:47042.

—— [1996a], *On a class of Hermitian kernels*; preprint.

—— [1996b], *Representations of Hermitian kernels by means of Kreĭn spaces*; preprint.

—— [1996c], *Kolmogorov decompositions and the realization of time-dependent systems*; preprint.

B. Ćurgus, A. Dijksma, H. Langer, and H. S. V. de Snoo [1989], *Characteristic functions of unitary colligations and of bounded operators in Kreĭn spaces*, Oper. Theory: Adv. Appl., vol. 41, Birkhäuser, Basel, pp. 125–152; MR 91c:47020.

B. Ćurgus and H. Langer [1990], *On a paper of de Branges*; preprint.

B. M. Davis and J. E. McCarthy [1991], *Multipliers of de Branges spaces*, Michigan Math. J. **38**, 225–240; MR 92k:46034.

Ch. Davis [1970], *J-unitary dilation of a general operator*, Acta Sci. Math. (Szeged) **31**, 75–86; MR 41#9032.

P. Delsarte and Y. Genin [1992], *On a generalization of the Szegő-Levinson recurrence and its application in lossless inverse scattering*, IEEE Trans. Inform. Theory **38**, no. 1, 104–110; MR 92i:94003.

P. Dewilde and H. Dym [1984], *Lossless inverse scattering, digital filters, and estimation theory*, IEEE Trans. Inform. Theory **30**, no. 4, 644–662; MR 85m:94012.

A. Dijksma and H. Langer [to appear], *Notes on a Nevanlinna-Pick interpolation problem for generalized Nevanlinna functions*, Topics in Interpolation Theory (Proceedings of a conference dedicated to the memory of V. P. Potapov), Oper. Theory: Adv. Appl., in preparation.

A. Dijksma, H. Langer, and H. S. V. de Snoo [1986a], *Characteristic functions of unitary operator colligations in Π_κ-spaces*, Oper. Theory: Adv. Appl., vol. 19, Birkhäuser, Basel, pp. 125–194; MR 89a:47053.

_____ [1986b], *Unitary colligations in Π_κ-spaces, characteristic functions and Štraus extensions*, Pacific J. Math. **125**, 347–362; MR 88e:47043.

_____ [1987a], *Unitary colligations in Kreĭn spaces and their role in the extension theory of isometries and symmetric linear relations in Hilbert spaces*, Lecture Notes in Math., vol. 1242, Functional Analysis, II (Dubrovnik, 1985), Springer-Verlag, Berlin-New York, pp. 1–42; MR 89a:47055.

_____ [1987b], *Representations of holomorphic operator functions by means of resolvents of unitary or selfadjoint operators in Kreĭn spaces*, Oper. Theory: Adv. Appl., vol. 24, Birkhäuser, Basel, pp. 123–143; MR 89a:47054.

A. Dijksma, S. A. M. Marcantognini, and H. S. V. de Snoo [1994], *A Schur type analysis of the minimal unitary Hilbert space extensions of a Kreĭn space isometry whose defect subspaces are Hilbert spaces*, Z. Anal. Anwendungen **13**, 233–260; MR 95f:47013.

W. F. Donoghue, Jr. [1974], *Monotone Matrix Functions and Analytic Continuation*, Grundlehren Math. Wiss., Bd. 207, Springer-Verlag, New York-Heidelberg; MR 58#6279.

M. A. Dritschel [1993], *The essential uniqueness property for linear operators in Kreĭn spaces*, J. Funct. Anal. **118**, 198–248; MR 94j:47051.

M. A. Dritschel and J. Rovnyak [1990], *Extension theorems for contraction operators on Kreĭn spaces*, Oper. Theory: Adv. Appl., vol. 47, Birkhäuser, Basel, pp. 221–305; MR 92m:47068.

_____ [1991], *Julia operators and complementation in Kreĭn spaces*, Indiana Univ. Math. J. **40**, 885–901; MR 93b:46042.

_____ [1996], *Operators on indefinite inner product spaces*, Lectures on Operator Theory and its Applications, Fields Institute Monographs, vol. 3, Amer. Math. Soc., Providence, RI; *Supplement to "Operators on indefinite inner product spaces,"* preprint.

V. K. Dubovoj, B. Fritzsche, and B. Kirstein [1992], *Matricial Version of the Classical Schur Problem*, Teubner-Texte zur Mathematik, vol. 129, B. G. Teubner Verlagsgesellschaft, Stuttgart; MR 93e:47021.

H. Dym [1989a], *J Contractive Matrix Functions, Reproducing Kernel Hilbert Spaces and Interpolation*, CBMS Regional Conference Series in Math., vol. 71, Amer. Math. Soc., Providence, RI; MR 90g:47003.

—— [1989b], *On reproducing kernel spaces, J unitary matrix functions, interpolation and displacement rank*, Oper. Theory: Adv. Appl., vol. 41, Birkhäuser, Basel, pp. 173–239; MR 92f:46024.

—— [1994a], *Shifts, realizations and interpolation, redux*, Oper. Theory: Adv. Appl., vol. 73, Birkhäuser, Basel, pp. 182–243; MR 96a:47029.

—— [1994b], *On the zeros of some continuous analogues of matrix orthogonal polynomials and a related extension problem with negative squares*, Comm. Pure Appl. Math. **47**, no. 2, 207–256; MR 95e:47018.

A. P. Filimonov [1977], *The inverse problem of the theory of characteristic operator functions of metric colligations in the space* Π_χ, Dokl. Akad. Nauk Ukrain. SSR Ser. A, no. 9, 798–802, 864; MR 58#12501.

P. A. Fuhrmann [1981], *Linear Systems and Operators in Hilbert Space*, McGraw-Hill, New York; MR 83d:47001.

Y. Genin, P. Van Dooren, T. Kailath, J.-M. Delosme, and M. Morf [1983], *On Σ-lossless transfer functions and related questions*, Linear Algebra Appl. **50**, 251–275; MR 85g:93024.

A. Gheondea [1993], *Quasi-contractions on Kreĭn spaces*, Oper. Theory: Adv. Appl., vol. 61, Birkhäuser, Basel, pp. 123–148; MR 94m:47070.

Ju. P. Ginzburg [1967a], *Multiplicative representations of J-contractive operator functions. I*, Mat. Issled. **2**, vyp. 2, 52–83; English transl.: Amer. Math. Soc. Transl. (2) **96** (1970), 189–221; MR 38#1551a.

—— [1967b], *Multiplicative representations of J-contractive operator functions. II*, Mat. Issled. **2**, vyp. 3, 20–51; English transl.: Amer. Math. Soc. Transl. (2) **96** (1970), 223–254; MR 38#1551b.

K. Glover [1984], *All optimal Hankel-norm approximations of linear multivariable systems and their L^∞-error bounds*, Internat. J. Control **39**, 1115–1193; MR 86a:93029.

L. B. Golinskiĭ [1983], *A generalization of the matrix Nevanlinna-Pick problem*, Izv. Akad. Nauk Armyan. SSR Ser. Mat. **18**, no. 3, 187–205; MR 85g:47049; English transl.: Soviet J. Contemporary Math. Anal. **18** (1983), no. 3, 22–39; MR 85g:47049.

J. Guyker [1991], *The de Branges-Rovnyak model*, Proc. Amer. Math. Soc. **111**, 95–99.; MR 91h:47009.

—— [1995], *The de Branges-Rovnyak model with finite-dimensional coefficients*, Trans. Amer. Math. Soc. **347**, 1383–1389; MR 95g:46045.

P. R. Halmos [1958], *Finite-Dimensional Vector Spaces*, 2-nd ed., Van Nostrand, New York; MR 4,11a and 19,725b.

_____ [1961], *Shifts on Hilbert spaces*, J. Reine Angew. Math. **208**, 102–112; MR 27#2868.

T. Hara [1992], *Operator inequalities and construction of Kreĭn spaces*, Integral Equations and Operator Theory **15**, 551–567; MR 93f:46027.

H. Helson [1964], *Lectures on Invariant Subspaces*, Academic Press, New York; MR 30#1409.

J. W. Helton [1974], *Discrete time systems, operator models, and scattering theory*, J. Functional Anal. **16**, 15–38; MR 56#3652.

I. S. Iokhvidov, M. G. Kreĭn, and H. Langer [1982], *Introduction to the Spectral Theory of Operators in Spaces with an Indefinite Metric*, Mathematical Research, vol. 9, Akademie-Verlag, Berlin; MR 85g:47050.

M. A. Kaashoek [1996], *State-space theory of rational matrix functions and applications*, Lectures on Operator Theory and its Applications, Fields Institute Monographs, vol. 3, Amer. Math. Soc., Providence, RI; MR 96i:15022.

T. Kailath [1980], *Linear Systems*, Prentice-Hall, Englewood Cliffs, NJ; MR 82a:93001.

T. Kailath and A. H. Sayed [1995], *Displacement structure: theory and applications*, SIAM Rev. **37**, no. 3, 297–386; MR 96h:15015.

V. E. Katsnelson and B. Kirstein [1997], *On the theory of matrix-valued functions belonging to the Smirnov class*, Topics in Interpolation Theory, eds. H. Dym, B. Fritsche, V. Katsnelson, B. Kirstein, Oper. Theory: Adv. Appl., Birkhäuser, Basel, pp. 299–350.

A. Ya. Kheifets and P. M. Yuditskiĭ [1994], *An analysis and extension of V. P. Potapov's approach to interpolation problems with applications to the generalized bi-tangential Schur-Nevanlinna-Pick problem and J-inner-outer factorization*, Oper. Theory: Adv. Appl., vol. 72, Birkhäuser, Basel, pp. 133–161; MR 95m:47022.

M. G. Kreĭn and H. Langer [1971ab], *The defect subspaces and generalized resolvents of an Hermitian operator in the space* Π_κ, Funkcional. Anal. i Priložen **5** (1971), no. 2, 59–71; ibid. **5** (1971), no. 3, 54–69; English transl.: Funct. Anal. Appl. **5** (1971), 136–146; ibid. **5** (1971), 217–228; MR 43#7951ab.

_____ [1972], *Über die verallgemeinerten Resolventen und die charakteristische Funktion eines isometrischen Operators im Raume* Π_κ, Hilbert Space Operators and Operator Algebras (Proc. Internat. Conf., Tihany, 1970); Colloq. Math. Soc. Janos Bolyai, vol. 5, North-Holland, Amsterdam, pp. 353–399; MR 54#11103.

_____ [1973], *Über die Q-Funktion eines* π-*hermiteschen Operators im Raume* Π_κ, Acta Sci. Math. (Szeged) **34**, 191–230; MR 47#7504.

—— [1977–1985], *Über einige Fortsetzungsprobleme, die eng mit der Theorie hermitescher Operatoren im Raume Π_κ zusammenhängen.*
I. *Einige Funktionenklassen und ihre Darstellungen*, Math. Nachr. **77** (1977), 187–236; MR 57#1173;
II. *Verallgemeinerte Resolventen, u-Resolventen und ganze Operatoren*, J. Funct. Anal. **30** (1978), 390–447; MR 80h:47045;
III. *Indefinite analogues of the Hamburger and Stieltjes moment problems. Part* I, Beiträge Anal. No. 14 (1979) 25–40; *Part* II, ibid. No. 15 (1980), 27–45 (1981); MR 83b:47047ab;
IV. *Continuous analogues of orthogonal polynomials on the unit circle with respect to an indefinite weight and related continuation problems for some classes of functions*, J. Operator Theory **13** (1985), 299–417; MR 87h:47084.

—— [1981], *Some propositions on analytic matrix functions related to the theory of operators in the space* Π_κ, Acta Sci. Math. (Szeged) **43**, 181–205; MR 82i:47053.

A. Kuzhel [1996], *Characteristic Functions and Models of Nonselfadjoint Operators*, Mathematics and its Applications, vol. 349, Kluwer, Dordrecht; MR 97e:47014.

R. B. Leech [1969], *On the characterization of $\mathcal{H}(B)$ spaces*, Proc. Amer. Math. Soc. **23**, 518–520; MR 40#3292.

M. S. Livšic, N. Kravitsky, A. S. Markus, and V. Vinnikov [1995], *Theory of commuting nonselfadjoint operators*, Mathematics and its Applications, vol. 332, Kluwer, Dordrecht; MR 96m:47003.

M. S. Livšic and V. P. Potapov [1950], *A theorem on the multiplication of characteristic matrix functions*, Dokl. Akad. Nauk SSSR (N.S.) **72**, 625–628; MR 11,669f.

B. A. Lotto and D. Sarason [1993], *Multipliers of de Branges-Rovnyak spaces*, Indiana Univ. Math. J. **42**, 907–920; MR 95a:46039.

S. A. M. Marcantognini [1990], *Unitary colligations of operators in Kreĭn spaces*, Integral Equations and Operator Theory **13**, 701–727; MR 92f:47010.

B. McEnnis [1979], *Purely contractive analytic functions and characteristic functions of noncontractions*, Acta Sci. Math. (Szeged) **41**, 161–172; MR 80i:47049.

M. Möller [1991], *Isometric and contractive operators in Kreĭn spaces*, Algebra i Analiz **3**, no. 3, 110–126; St. Petersburg Math. J. **3** (1992), no. 3, 595–611; MR 93a:47037.

N. K. Nikol'skiĭ and V. I. Vasyunin [1986], *Notes on two function models*, The Bieberbach Conjecture (West Lafayette, Ind., 1985), Math. Surveys Monographs, vol. 21, Amer. Math. Soc., Providence, RI, pp. 113–141; MR 88f:47008.

_____ [1989], *A unified approach to function models, and the transcription problem*, Oper. Theory: Adv. Appl., vol. 41, Birkhäuser, Basel, pp. 405–434; MR 91c:47017.

V. P. Potapov [1955], *The multiplicative structure of J-contractive matrix functions*, Trudy Moskov. Mat. Obšč. **4**, 125–236; English transl.: Amer. Math. Soc. Transl. (2) **15** (1960), 131–243; MR 17,958f.

V. Pták and P. Vrbová [1993], *An abstract analogon of the de Branges-Rovnyak functional model*, Integral Equations and Operator Theory **16**, 565–599; MR 94c:47010.

M. Rosenblum and J. Rovnyak [1982], *An operator-theoretic approach to theorems of the Pick-Nevanlinna and Loewner types. II*, Integral Equations and Operator Theory **5**, 870–887; MR 84b:47020.

_____ [1985], *Hardy Classes and Operator Theory*, Oxford Mathematical Monographs, Oxford University Press, New York; Dover republication, New York, 1997; MR 87e:47001.

_____ [1994], *Topics in Hardy Classes and Univalent Functions*, Birkhäuser Advanced Texts: Basler Lehrbücher, Birkhäuser, Basel; MR 97a:30047.

J. Rovnyak [1968], *Characterization of spaces $\mathfrak{H}(M)$*, unpublished paper.

L. A. Sakhnovich [1986], *Problems of factorization and operator identities*, Uspekhi Mat. Nauk **41**, no. 1(247), 4–55, 240; English transl.: Russian Math. Surveys **41** (1986), no. 1, 1–64; MR 87k:47041.

M. Sand [1995], *Spaces contractively invariant for the backward shift*, J. Operator Theory **34**, 125–144; MR 96j:47025.

D. Sarason [1986a], *Shift-invariant spaces from the Brangesian point of view*, The Bieberbach Conjecture (West Lafayette, Ind., 1985), Math. Surveys Monographs, vol. 21, Amer. Math. Soc., Providence, RI, pp. 153–166; MR 88d:47014a.

_____ [1986b], *Doubly shift-invariant spaces in H^2*, J. Operator Theory **16**, 75–97; MR 88d:47014b.

_____ [1990], *Function theory and de Branges's spaces*, Operator Theory: Operator Algebras and Applications, Part 1 (Durham, NH, 1988), Proc. Sympos. Pure Math., vol. 51, Part 1, Amer. Math. Soc., Providence, RI, pp. 495–502.

_____ [1994], *Sub-Hardy Hilbert Spaces in the Unit Disk*, University of Arkansas Lectures Notes in the Mathematical Sciences, vol. 10, Wiley, New York; MR 96k:46039.

I. Schur [1917,1918], *Über Potenzreihen, die im Innern des Einheitskreises beschränkt sind. I*, J. Reine Angew. Math. **147** (1917), 205–232; II, ibid. **148** (1918), 122–145; Gesammelte Abhandlungen, bd. II, nos. 29, 30 (pp. 137–188).

L. Schwartz [1964], *Sous-espaces hilbertiens d'espaces vectoriels topologiques et noyaux associés (noyaux reproduisants)*, J. Analyse Math. **13**, 115–256; MR 31#3835.

Yu. L. Shmul'yan [1967], *Division in the class of J-expansive operators*, Mat. Sb. (N.S.) **74** (116), 516–525; English transl.: Math. USSR-Sbornik **3** (1967), 471–479; MR 37#784.

P. Sorjonen [1975], *Pontrjaginräume mit einem reproduzierenden Kern*, Ann. Acad. Sci. Fenn. Ser. A I Math., no. 594, 30 pages; MR 53#8875.

A. V. Štraus [1960], *Characteristic functions of linear operators*, Izv. Akad. Nauk SSSR Ser. Mat. **24**, 43–74; English transl.: Amer. Math. Soc. Transl. (2) **40** (1964), 1–37; MR 25#4363.

—— [1965], *Extensions of a symmetric operator depending on a parameter*, Izv. Akad. Nauk SSSR Ser. Mat. **29**, 1389–1416; English transl.: Amer. Math. Soc. Transl. (2) **61** (1967), 113–141; MR 34#3338.

—— [1966], *One-parameter families of extensions of a symmetric operator*, Izv. Akad. Nauk SSSR Ser. Mat. **30**, 1325–1352; English transl.: Amer. Math. Soc. Transl. (2) **90** (1970), 135–164; MR 34#3339.

—— [1968], *Extensions and characteristic function of a symmetric operator*, Izv. Akad. Nauk SSSR Ser. Mat. **32**, 186–207; English transl.: Math. USSR Izv. **2** (1968), 181–204; MR 37#788.

B. Sz.-Nagy and C. Foiaş [1970], *Harmonic Analysis of Operators on Hilbert Space*, North-Holland, New York; MR 43#947.

B. Sz.-Nagy and A. Korányi [1958], *Operatortheoretische Behandlung und Verallgemeinerung eines Problemkreises in der komplexen Funktionentheorie*, Acta Math. **100**, 171–202; MR 24#A437.

A. M. Yang [1994], *A construction of unitary linear systems*, Integral Equations and Operator Theory **19**, 477–499; MR 95i:47019.

NOTATION INDEX

Special colligations

AUTHOR INDEX

SUBJECT INDEX

Titles previously published in the series

OPERATOR THEORY: ADVANCES AND APPLICATIONS
BIRKHÄUSER VERLAG

Edited by
I. Gohberg,
School of Mathematical Sciences, Tel-Aviv University, Ramat Aviv, Israel

This series is devoted to the publication of current research in operator theory, with particular emphasis on applications to classical analysis and the theory of integral equations, as well as to numerical analysis, mathematical physics and mathematical methods in electrical engineering.

58. **I. Gohberg** (Ed.): Continuous and Discrete Fourier Transforms, Extension Problems and Wiener-Hopf Equations, 1992, (3-7643-2809-6)

59. **T. Ando, I. Gohberg** (Eds.): Operator Theory and Complex Analysis, 1992, (3-7643-2824-X)

60. **P.A. Kuchment:** Floquet Theory for Partial Differential Equations, 1993, (3-7643-2901-7)

61. **A. Gheondea, D. Timotin, F.-H. Vasilescu** (Eds.): Operator Extensions, Interpolation of Functions and Related Topics, 1993, (3-7643-2902-5)

62. **T. Furuta, I. Gohberg, T. Nakazi** (Eds.): Contributions to Operator Theory and its Applications. The Tsuyoshi Ando Anniversary Volume, 1993, (3-7643-2928-9)

63. **I. Gohberg, S. Goldberg, M.A. Kaashoek:** Classes of Linear Operators, Volume 2, 1993, (3-7643-2944-0)

64. **I. Gohberg** (Ed.): New Aspects in Interpolation and Completion Theories, 1993, (3-7643-2948-3)

65. **M.M. Djrbashian:** Harmonic Analysis and Boundary Value Problems in the Complex Domain, 1993, (3-7643-2855-X)

66. **V. Khatskevich, D. Shoiykhet:** Differentiable Operators and Nonlinear Equations, 1993, (3-7643-2929-7)

67. **N.V. Govorov †:** Riemann's Boundary Problem with Infinite Index, 1994, (3-7643-2999-8)

68. **A. Halanay, V. Ionescu:** Time-Varying Discrete Linear Systems Input-Output Operators. Riccati Equations. Disturbance Attenuation, 1994, (3-7643-5012-1)

69. **A. Ashyralyev, P.E. Sobolevskii:** Well-Posedness of Parabolic Difference Equations, 1994, (3-7643-5024-5)

70. **M. Demuth, P. Exner, G. Neidhardt, V. Zagrebnov** (Eds): Mathematical Results in Quantum Mechanics. International Conference in Blossin (Germany), May 17–21, 1993, 1994, (3-7643-5025-3)

71. **E.L. Basor, I. Gohberg** (Eds): Toeplitz Operators and Related Topics. The Harold Widom Anniversary Volume. Workshop on Toeplitz and Wiener-Hopf Operators, Santa Cruz, California, September 20–22, 1992, 1994 (3-7643-5068-7)

72. **I. Gohberg, L.A. Sakhnovich** (Eds): Matrix and Operator Valued Functions. The Vladimir Petrovich Potapov Memorial Volume, (3-7643-5091-1)

73. **A. Feintuch, I. Gohberg** (Eds): Nonselfadjoint Operators and Related Topics. Workshop on Operator Theory and Its Applications, Beersheva, February 24–28, 1994, (3-7643-5097-0)

74. **R. Hagen, S. Roch, B. Silbermann:** Spectral Theory of Approximation Methods for Convolution Equations, 1994, (3-7643-5112-8)

75. **C.B. Huijsmans, M.A. Kaashoek, B. de Pagter**: Operator Theory in Function Spaces and Banach Lattices. The A.C. Zaanen Anniversary Volume, 1994 (ISBN 3-7643-5146-2)

76. **A.M. Krasnosellskii:** Asymptotics of Nonlinearities and Operator Equations, 1995, (ISBN 3-7643-5175-6)

77. **J. Lindenstrauss, V.D. Milman** (Eds): Geometric Aspects of Functional Analysis Israel Seminar GAFA 1992–94, 1995, (ISBN 3-7643-5207-8)

78. **M. Demuth, B.-W. Schulze** (Eds): Partial Differential Operators and Mathematical Physics: International Conference in Holzhau (Germany), July 3–9, 1994, 1995, (ISBN 3-7643-5208-6)

79. **I. Gohberg, M.A. Kaashoek, F. van Schagen**: Partially Specified Matrices and Operators: Classification, Completion, Applications, 1995, (ISBN 3-7643-5259-0)

80. **I. Gohberg, H. Langer** (Eds): Operator Theory and Boundary Eigenvalue Problems. International Workshop in Vienna, July 27–30, 1993, 1995, (ISBN 3-7643-5275-2)

81. **H. Upmeier**: Toeplitz Operators and Index Theory in Several Complex Variables, 1996, (ISBN 3-7643-5282-5)

82. **T. Constantinescu**: Schur Parameters, Factorization and Dilation Problems, 1996, (ISBN 3-7643-5285-X)

83. **A.B. Antonevich**: Linear Functional Equations. Operator Approach, 1995, (ISBN 3-7643-2931-9)

84. **L.A. Sakhnovich**: Integral Equations with Difference Kernels on Finite Intervals, 1996, (ISBN 3-7643-5267-1)

85/ **Y.M. Berezansky, G.F. Us, Z.G. Sheftel**: Functional Analysis, Vol. I + Vol. II, 1996,
86. Vol. I (ISBN 3-7643-5344-9), Vol. II (3-7643-5345-7)

87. **I. Gohberg, P. Lancaster, P.N. Shivakumar** (Eds): Recent Developments in Operator Theory and Its Applications. International Conference in Winnipeg, October 2–6, 1994, 1996, (ISBN 3-7643-5414-5)

88. **J. van Neerven** (Ed.): The Asymptotic Behaviour of Semigroups of Linear Operators, 1996, (ISBN 3-7643-5455-0)

89. **Y. Egorov, V. Kondratiev**: On Spectral Theory of Elliptic Operators, 1996, (ISBN 3-7643-5390-2)

90. **A. Böttcher, I. Gohberg** (Eds): Singular Integral Operators and Related Topics. Joint German-Israeli Workshop, Tel Aviv, March 1–10, 1995, 1996, (ISBN 3-7643-5466-6)

91. **A.L. Skubachevskii**: Elliptic Functional Differential Equations and Applications, 1997, (ISBN 3-7643-5404-6)

92. **A.Ya. Shklyar**: Complete Second Order Linear Differential Equations in Hilbert Spaces, 1997, (ISBN 3-7643-5377-5)

93. **Y. Egorov, B.-W. Schulze**: Pseudo-Differential Operators, Singularities, Applications, 1997, (ISBN 3-7643-5484-4)

94. **M.I. Kadets, V.M. Kadets**: Series in Banach Spaces. Conditional and Unconditional Convergence, 1997, (ISBN 3-7643-5401-1)

95. **H. Dym, V. Katsnelson, B. Fritzsche, B. Kirstein** (Eds): Topics in Interpolation Theory, 1997. (ISBN 3-7643-5723-1)

IEOT

Integral Equations and Operator Thoery

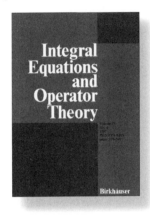

ISSN 0378-620X

Aims and Scope

Integral Equations and Operator Theory (IEOT) appears monthly and is devoted to the publication of current research in integral equations, operator theory and related topics with emphasis on the linear aspects of the theory. The journal reports on the full scope of curent developments from abstract theory to numerical methods and applications to analysis, physics, mechanics, engineering and others. The journal consists of two sections: a main section consisting of refereed papers and a second consisting of short announcements of important results, open problems, information, etc.

Abstracted/Indexed in:
CompuMath Citation Index, Current Contents,
Mathematical Reviews, Zentralblatt für Mathematik,
Mathematics Abstracts, DB MATH

Subscription Information for 1997
IEOT is published in 3 volumes per year,
and 4 issues per volume
Volumes 27 - 29 (1997)

Founded in 1978 by Birkhäuser Verlag AG
3 volumes per year, 4 issues per volume
approx. 500 pages per volume
Format: 17 x 24 cm
Back volumes are available

For orders originating from all over
the world except USA and Canada:
Birkhäuser Verlag AG
P.O Box 133
CH-4010 Basel/Switzerland
Fax: +41/61/205 07 92
e-mail: farnik@birkhauser.ch

For orders originating in the
USA and Canada:
Birkhäuser
333 Meadowland Parkway
USA-Secaurus, NJ 07094-2491
Fax: +1 201 348 4033
e-mail: orders@birkhauser.com

Birkhäuser

Birkhäuser Verlag AG
Basel · Boston · Berlin

VISIT OUR HOMEPAGE **http://www.birkhauser.ch**

Mathematics with Birkhäuser

BAT
Birkhäuser Advanced Texts / Basler Lehrbücher

M. Rosenblum / J. Rovnyak,
University of Virginia, Charlottesville, VA, USA

Topics in Hardy Classes and Univalent Functions

1994. 264 pages. Hardcover
ISBN 3-7643-5111-X

This book treats classical and contemporary topics in function theory and is accessible after a one-year course in real and complex analysis. It can be used as a text for topics courses or read independently by graduate students and researchers in function theory, operator theory, and applied areas.

The first six chapters supplement the authors' book, Hardy Classes and Operator Theory. The theory of harmonic majorants for subharmonic functions is used to introduce Hardy-Orlicz classes, which are specialized to standard Hardy classes on the unit disk. The theorem of Szegö-Solomentsev characterizes boundary behavior. Half-plane function theory receives equal treatment and features the theorem of Flett and Kuran on existence of harmonic majorants and applications of the Phragmén-Lindelöf principle.

The last three chapters contain an introduction to univalent functions, leading to a self-contained account of Loewner's differential equation and de Branges' proof of the Milin conjecture.

"...I can think of no authors better able than the present ones to give an exposition of the modern theory of univalent functions and they have succeeded admirably...One only hopes that, when it is written, it is as well written as the present volume...The book is written for graduate students. It is a compelling introduction to this fascinating subject and is warmly recommended."

J.M. Anderson, Proceedings of the Edinburgh Mathematical Society, 1995/38

For orders originating from all over the world except USA and Canada:
Birkhäuser Verlag AG
P.O Box 133
CH-4010 Basel/Switzerland
Fax: +41/61/205 07 92
e-mail: farnik@birkhauser.ch

For orders originating in the USA and Canada:
Birkhäuser
333 Meadowland Parkway
USA-Secarus, NJ 07094-2491
Fax: +1 201 348 4033
e-mail: orders@birkhauser.com

Birkhäuser
Birkhäuser Verlag AG
Basel · Boston · Berlin

VISIT OUR HOMEPAGE **http://www.birkhauser.ch**

Mathematics with Birkhäuser

ANALYSIS • GEOMETRY • LINEAR ALGEBRA

ISNM 123 • International Series of Numerical Mathematics

C. Bandle, University of Basel, Switzerland /
W.N. Everitt, University of Birmingham, UK /
L. Losonczi, Kuwait University, Kuweit /
Walter, University of Karlsruhe, Germany (Eds)

General Inequalities 7

**7th International Conference,
Oberwolfach, November 13–18, 1995**

1997. Approx. 416 pages. Hardcover
ISBN 3-7643-5722-3

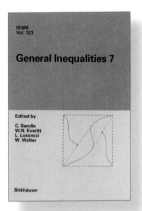

This is perhaps the last remaining field comprehended and used by mathematicians in all areas of the discipline. New inequalities are discovered every year, some for their intrinsic interest whilst others flow from results obtained in various branches of mathematics. The study of inequalities reflects the many and various aspects of mathematics. There are numerous applications in a wide variety of fields, from mathematical physics to biology and economics. This volume contains the proceedings of the General Inequalities 7 meeting held at Oberwolfach in November 1995. It is conceived in the spirit of the preceding volumes in the sense that it not only contains the latest results, but it is also a useful reference for lecturers and researchers.

For orders originating from all over
the world except USA and Canada:
Birkhäuser Verlag AG
P.O Box 133
CH-4010 Basel/Switzerland
Fax: +41/61/205 07 92
e-mail: farnik@birkhauser.ch

For orders originating in the
USA and Canada:
Birkhäuser
333 Meadowland Parkway
USA-Secaurus, NJ 07094-2491
Fax: +1 201 348 4033
e-mail: orders@birkhauser.com

Birkhäuser

**Birkhäuser Verlag AG
Basel · Boston · Berlin**

VISIT OUR HOMEPAGE **http://www.birkhauser.ch**